SpringerBriefs in Modern Perspectives on Disability Research

This book series on disability research is a comprehensive collection of research on disability and related issues. The series is designed to promote interdisciplinary collaboration and exchange, bringing together scholars and practitioners from different fields to share their perspectives and insights. Disability research is an interdisciplinary field that examines the social, cultural, historical, and political dimensions of disability. It encompasses a wide range of topics, including disability rights, accessibility, assistive technologies, healthcare, education, employment, and social welfare. Disability research scholars employ a range of theoretical and methodological approaches to understand the experiences of people with disabilities, as well as the ways in which disability intersects with other social identities such as race, gender, sexuality, and class.

The series seeks to advance knowledge and understanding of disability by publishing rigorous, innovative, and relevant research. It aims to promote disability rights and social justice by highlighting the ways in which people with disabilities are marginalized and discriminated against in society, and advocating for greater social inclusion and accessibility. The series also seeks to inform policy and practice by disseminating research findings that can help to shape policy decisions and contribute to positive social change.

Moumita Kundu • Subhayan Das
Editors

Alzheimer's Disease-Associated Cognitive Disabilities

Springer

Editors
Moumita Kundu
Department of Pharmaceutical Technology
Brainware University
Barasat, West Bengal, India

Subhayan Das
Department of Allied Health Sciences
Brainware University
Barasat, West Bengal, India

ISSN 3004-9709 ISSN 3004-9717 (electronic)
SpringerBriefs in Modern Perspectives on Disability Research
ISBN 978-981-95-7120-8 ISBN 978-981-95-7121-5 (eBook)
https://doi.org/10.1007/978-981-95-7121-5

This Springer imprint is published by the registered company Springer Nature Singapore Pte Ltd.
The registered company address is: 152 Beach Road, #21-01/04 Gateway East, Singapore 189721, Singapore

Contents

Contents

Chapter 1
Introduction: Stages and Progression of Alzheimer's Disease

Shayeri Chatterjee Ganguly and Biraj Patra

Abstract Alzheimer's disease (AD) is an irreversible, degenerative, and progressive neuronal disease leading to moderate memory impairment during its initial stages. The condition progressively damages memory and cognitive functions, resulting in diminished reaction of the patient toward the external atmosphere in the latter stages. According to the 2016 World Alzheimer Report, roughly 46.8 million individuals are globally afflicted with AD. The onset of symptoms is categorized as early-onset or late-onset AD. Patients with early onset typically have a non-memory phenotype characterized by visuospatial deficits. Early-onset autosomal dominant AD is linked to amyloid-β (Aβ) irregularities. Dementia is the chief characteristic of late-onset type. A recent study indicates that late-onset AD is characterized as a polygenic condition involving unusual interactions among various biological pathways. The primary risk factor for AD is age, succeeded by the cardiovascular and lifestyle-related risk factors, along with alterations that occur in the blood–brain barrier and cerebral vasculature. This chapter provides a comprehensive overview of the stages and progression of AD, emphasizing the biological, clinical, and epidemiological dimensions of the disease.

S. C. Ganguly (✉) · B. Patra
Department of Pharmaceutical Technology, Brainware University, Barasat, Kolkata, West Bengal, India
e-mail: shayeri.ch1992@gmail.com

M. Kundu, S. Das (eds.), *Alzheimer's Disease-Associated Cognitive Disabilities*, SpringerBriefs in Modern Perspectives on Disability Research,
https://doi.org/10.1007/978-981-95-7121-5_1

Graphical Abstract

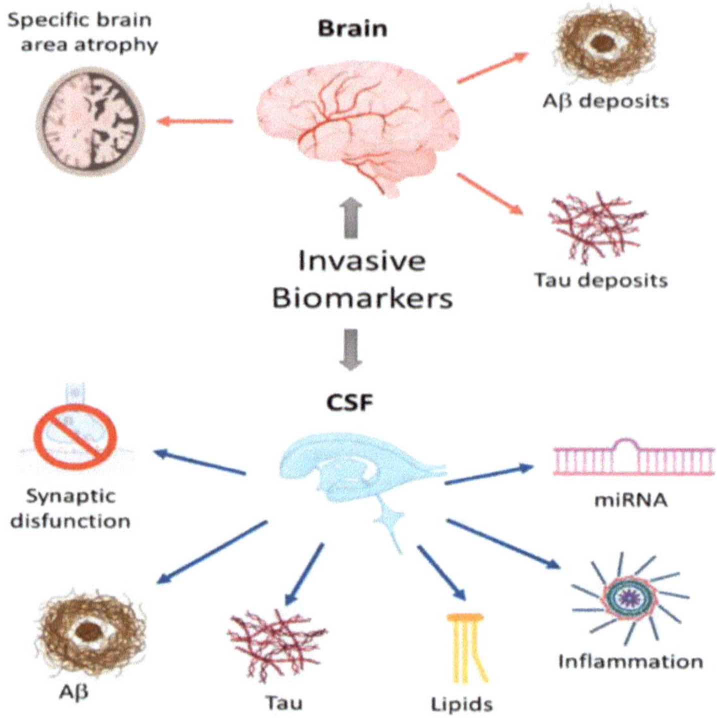

Keywords Degenerative · Dementia · Polygenic · Blood–brain barrier · Cerebral vasculature

1.1 Introduction

Alzheimer's disease (AD) is the most common cause of dementia worldwide, representing 60–70% of all cases. It is a progressive, irreversible neurodegenerative disorder that slowly erodes memory, cognitive abilities, and functional independence. The disease has far-reaching implications not only for affected individuals but also for families, healthcare systems, and societies at large (Arvanitakis et al., 2019). As populations age globally, the prevalence of AD is increasing at an alarming rate. Estimates suggest that around 55 million people were living with dementia in 2020, and this number is projected to rise to nearly 139 million by 2050 (Livingston et al., 2017, 2020). The socioeconomic burden is equally staggering, with costs of dementia care surpassing $1 trillion annually and expected to double within the next two decades. These costs encompass direct medical care, social support services, and the informal caregiving provided by families, which often comes at immense personal and financial sacrifice (Heneka et al., 2015a, 2015b; Zetterberg & Blennow, 2021).

From a clinical perspective, AD is insidious in onset and slowly progressive in course. Symptoms typically begin with subtle memory lapses, such as forgetting recent conversations or misplacing objects, which are often mistaken for normal aging (Barabâi et al., 2011). Over time, the disease expands its footprint to impair language, executive function, visuospatial abilities, and ultimately the capacity to carry out basic activities of daily living. Behavioral and psychological symptoms such as apathy, agitation, depression, and hallucinations may also emerge, further complicating care (Arvanitakis et al., 2019; Brookmeyer et al., 2007). Ultimately, individuals in advanced stages require full-time support and supervision, making AD not just a medical condition but a profound social and caregiving challenge (Villemagne et al., 2013). Despite being identified over a century ago by Alois Alzheimer, the disease remains without a definitive cure. Traditional views once regarded AD as a single, monolithic entity that progressed in a linear fashion. However, decades of research have revealed that the disease is highly heterogeneous, with complex interactions between genetic predisposition, environmental influences, and lifestyle factors shaping its onset and progression (Cheng et al., 2012; Lin et al., 2016). The classic hallmarks—amyloid plaques and neurofibrillary tangles—are now understood to represent only part of a larger cascade that involves synaptic dysfunction, chronic inflammation, mitochondrial impairment, vascular compromise, and widespread neuronal loss (Braak & Braak, 1991; Mintun et al., 2021; Thal et al., 2002).

One of the most important conceptual advances in recent decades is the recognition that AD begins many years, even decades, before clinical symptoms appear (Donohue et al., 2014). This long preclinical phase is characterized by the silent accumulation of amyloid-β and tau pathology in the brain, detectable only through specialized biomarkers such as cerebrospinal fluid assays or positron emission tomography (PET) imaging (Guo & Lee, 2014). The gradual progression from silent pathology to mild cognitive impairment (MCI) and eventually dementia underscores the need for staging frameworks that can capture the disease across its entire spectrum. Such frameworks are essential for accurate diagnosis, prognostication, clinical trial design, and therapeutic intervention (Butterfield & Halliwell, 2019).

1.2 Clinical Stages of Alzheimer's Disease

1.2.1 Normal Cognition and Subjective Cognitive Decline

The Alzheimer's continuum begins in a stage where cognition is clinically normal. Individuals may function independently without measurable impairment. However, some people report subtle concerns about their memory or cognitive function, even though neuropsychological testing remains within normal limits (Roberts et al., 2000).

Table 1.1 Global burden of AD: prevalence, projections, costs, and disability impact

Metric	Value
Global prevalence (2020)	~55 million
Projected prevalence (2050)	~139 million
Annual economic cost	>$1 trillion
Years of life lost (DALYs)	Third leading cause in elderly

This condition, known as subjective cognitive decline (SCD), has gained attention as a possible earliest symptomatic manifestation of AD (Hansson et al., 2006). Not all cases of SCD represent incipient AD. Memory complaints are common in aging and can be influenced by anxiety, depression, or stress (Jessen et al., 2014). Nonetheless, when concerns are persistent and corroborated by informants, and especially when combined with risk factors such as family history or genetic predisposition, SCD is associated with an elevated likelihood of progression (Mitchell & Shiri-Feshki, 2009). Identifying individuals in this stage offers an opportunity for early counseling, risk modification, and possibly future preventative interventions. Table 1.1 represents the global burden of AD.

1.2.2 Normal Cognition and Subjective Cognitive Decline

The Alzheimer's continuum begins in a stage where cognition is clinically normal. Individuals may function independently without measurable impairment. However, some people report subtle concerns about their memory or cognitive function, even though neuropsychological testing remains within normal limits (Crutch et al., 2017). This condition, known as subjective cognitive decline (SCD), has gained attention as a possible earliest symptomatic manifestation of AD. Not all cases of SCD represent incipient AD. Memory complaints are common in aging and can be influenced by anxiety, depression, or stress. Nonetheless, when concerns are persistent and corroborated by informants, and especially when combined with risk factors such as family history or genetic predisposition, SCD is associated with an elevated likelihood of progression. Identifying individuals in this stage offers an opportunity for early counseling, risk modification, and possibly future preventative interventions (Sweeney et al., 2018).

1.2.3 Mild Cognitive Impairment (MCI)

Mild cognitive impairment represents the prodromal phase of AD (Albert et al., 2011). Individuals with MCI demonstrate measurable deficits, most often in episodic memory, though attention, language, or visuospatial domains can also be affected.

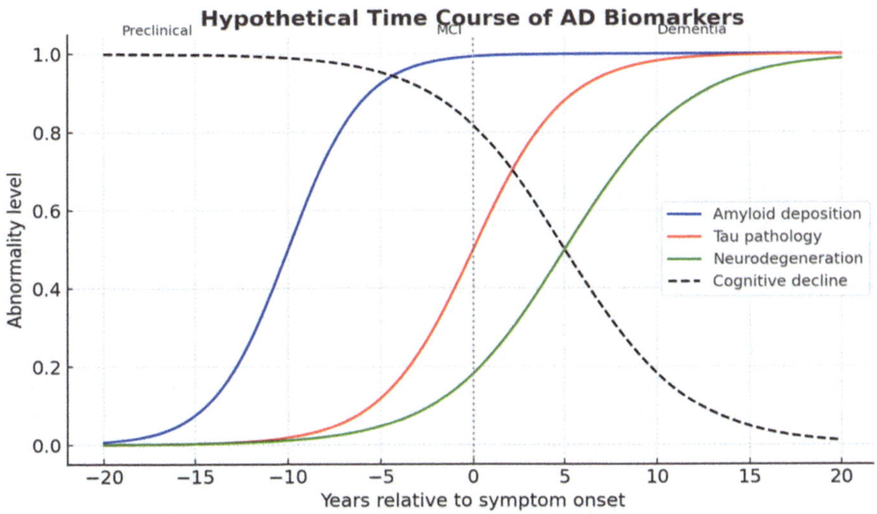

Fig. 1.1 Hypothetical time course of AD biomarkers, showing early amyloid deposition, tau pathology, neurodegeneration, and cognitive decline

Crucially, despite cognitive decline, individuals maintain independence in daily activities, distinguishing MCI from dementia. The annual progression rate from MCI to AD dementia varies, but it is substantially higher than in normal aging. Some individuals remain stable or even revert to normal cognition, but the presence of AD-related biomarkers dramatically increases the risk of conversion. In this sense, MCI due to AD can be conceptualized as a transitional but unstable state in the disease continuum (Iadecola et al., 2013). Figure 1.1 illustrates hypothetical time course of AD biomarkers.

1.2.4 Mild Alzheimer's Dementia

When cognitive deficits begin interfering with daily activities, the threshold of dementia is crossed. Mild Alzheimer's dementia is characterized by memory loss that disrupts daily functioning, difficulty learning new information, and frequent repetition. Patients may forget recent conversations, misplace belongings, or struggle with navigation in familiar areas (Folstein et al., 1975). Language difficulties, subtle personality changes, and impaired judgment may appear. This stage is often when families first seek medical attention, as the decline becomes noticeable in professional or social settings. Diagnosis at this stage relies on clinical evaluation, cognitive testing, and, increasingly, biomarker evidence supporting the underlying pathology (Petersen et al., 1999).

1.2.5 Moderate and Severe Alzheimer's Dementia

As the disease advances to the moderate stage, impairments become more widespread. Patients require help with personal care such as dressing, bathing, or managing finances. Memory deficits worsen, and patients may lose awareness of current events or even their own personal history. Language impairment becomes more pronounced, with reduced vocabulary, difficulty constructing sentences, or reliance on repetitive phrases (Lehmann et al., 2011; Lyketsos et al., 2011). Behavioral and psychological symptoms frequently emerge in this stage. These can include depression, anxiety, apathy, agitation, aggression, paranoia, or hallucinations. Wandering behavior and disorientation are common and may lead to safety concerns. Caregiver burden intensifies, and institutional support is often required. The severe stage of AD is characterized by profound cognitive decline and near-total dependence. Patients may lose the ability to recognize close family members or to communicate meaningfully. Mobility declines, leading to bed confinement. Swallowing difficulties increase the risk of aspiration pneumonia, while incontinence, malnutrition, and recurrent infections become common. Ultimately, AD becomes a systemic illness with multi-organ complications. The average duration from diagnosis to death is approximately 8–10 years, though this varies widely based on age, comorbidities, and genetic factors (Brookmeyer et al., 2002). Death typically results from secondary complications such as infections rather than from the neurodegenerative process itself (Heneka et al., 2015a; Sachdev et al., 2014). Table 1.2 depicts clinical symptom progression across various stages of AD.

1.2.6 Atypical Clinical Presentations

Although the most recognized presentation of AD is memory impairment, atypical forms highlight the disease's clinical heterogeneity (Mathys et al., 2019).

Table 1.2 Clinical symptom progression across stages of AD

Stage	Core symptoms	Support needs
Preclinical	No overt symptoms	None
Mild cognitive impairment (MCI)	Mild memory lapses, word-finding difficulty	Minimal
Mild dementia	Noticeable memory loss, confusion in daily tasks	Occasional help
Moderate dementia	Loss of independence, language/behavior changes	Daily assistance
Severe dementia	Severe cognitive & physical impairment	Full-time care

Posterior cortical atrophy is an atypical variant characterized by visuospatial difficulties, object recognition problems, and impaired reading. Logopenic variant primary progressive aphasia manifests with language deficits, particularly in word retrieval and repetition. A frontal variant can present with behavioral and executive dysfunction, often mimicking frontotemporal dementia (Crutch et al., 2017). Recognizing these atypical forms is vital for accurate diagnosis, staging, and treatment planning (Mesulam et al., 2014; Petersen et al., 1999).

1.2.7 Neuropathological Staging: Braak and Beyond

The pathology of AD is defined by two hallmark lesions: extracellular amyloid-β plaques and intracellular neurofibrillary tangles composed of hyperphosphorylated tau. Neuropathological staging systems, derived from postmortem studies, have elucidated how these lesions develop and spread throughout the brain (Janelidze et al., 2020; Palmqvist et al., 2020). The Braak and Braak system remains the cornerstone for describing the progression of tau pathology (Schöll et al., 2016). The earliest tangles appear in the transentorhinal region (stages I–II), then extend to limbic areas such as the hippocampus (stages III–IV), before spreading widely to neocortical association areas (stages V–VI) (Klunk et al., 2015). This spread corresponds with the clinical evolution of symptoms: memory impairment emerges as the hippocampus is affected, while widespread cognitive and functional deficits appear as neocortical regions are compromised (Mattsson et al., 2019). Figure 1.2 illustrates the Braak staging of tau pathology.

1.2.8 Amyloid Deposition

Amyloid plaques begin accumulating decades before symptoms, first in neocortical areas and later spreading to subcortical regions. Interestingly, the distribution of amyloid burden does not correlate as strongly with clinical severity as tau pathology does. Many cognitively normal older adults have significant amyloid deposition, underscoring its role as an initiating event rather than a direct driver of cognitive decline (Landau & Jagust, 2015). Figure 1.3 states the comparative progression between tau and amyloid pathology.

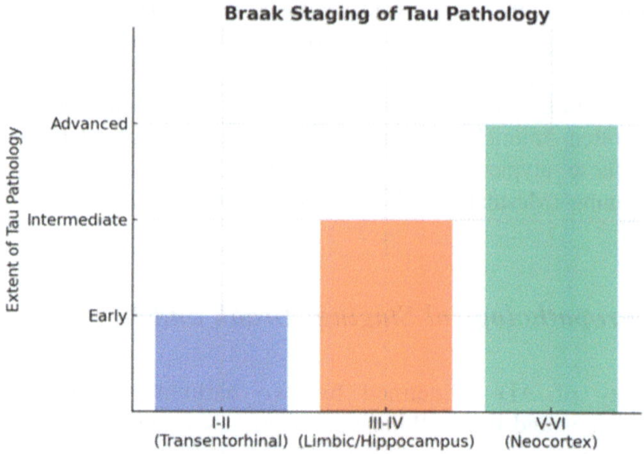

Fig. 1.2 Braak staging schematic for tau pathology progression

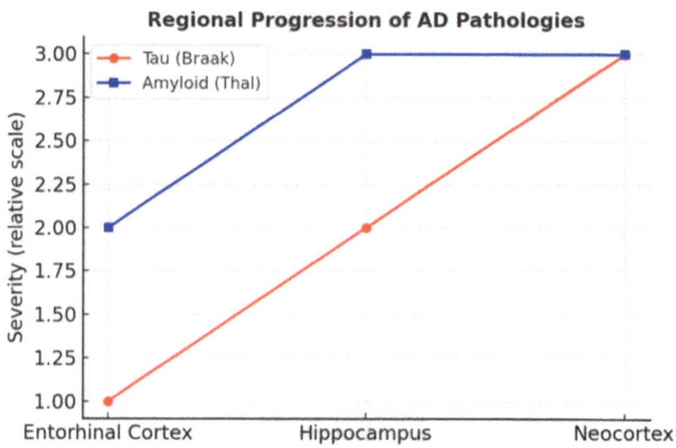

Fig. 1.3 Comparative progression of tau (Braak) and amyloid (Thal) pathology across brain regions

1.2.9 Additional Pathologies

Other pathologies frequently coexist with AD. TDP-43 inclusions, vascular brain injury, and Lewy body pathology are common in older adults and may accelerate decline or alter clinical presentations. The concept of "mixed dementia" reflects the frequent overlap of pathologies, complicating both staging and prognosis (Bennett et al., 2012a, 2012b; Kapasi et al., 2017; Irwin et al., 2013).

1.3 Limitations of Postmortem Staging

While neuropathological staging has provided fundamental insights, it is inherently retrospective. Translating Braak and amyloid staging into in vivo frameworks has required the development of biomarkers such as PET imaging and CSF assays (Johnson et al., 2016; Klunk et al., 2015). These technologies now allow clinicians and researchers to approximate neuropathological staging in living individuals, enabling earlier diagnosis and real-time monitoring.

1.4 Biomarker-Based Frameworks and Biological Staging

The recognition that AD pathology begins long before clinical symptoms has driven a shift toward biological definitions of disease (Ashley, 2016). Biomarker-based staging frameworks integrate molecular and imaging data to classify individuals along the AD continuum (Jack et al., 2018). The AT(N) system has become the standard for biomarker-based classification. "A" represents amyloid pathology, measured via amyloid PET or reduced CSF Aβ42 levels (Lewczuk et al., 2015). "T" represents tau pathology, measured through tau PET or CSF phosphorylated tau (Hampel et al., 2019). "N" represents neurodegeneration, assessed by MRI brain atrophy, Fluorodeoxyglucose (FDG)-PET hypometabolism, or neurofilament light levels in CSF or plasma. This framework decouples diagnosis from clinical symptoms (Talbot et al., 2012). An individual who is cognitively normal but has $A+T+$ biomarker status can be said to have biologically defined AD, even before dementia develops (Johnson et al., 2020). This has profound implications for early detection, risk stratification, and trial enrollment (Jack et al., 2016). As new biomarkers emerge, the AT(N) framework has been expanded (Kester et al., 2015). Markers of neuroinflammation, synaptic integrity, and vascular injury are increasingly integrated. For instance, microglial activation can be measured with PET ligands targeting the translocator protein, while synaptic dysfunction can be assessed through neurogranin levels in CSF (Mattsson-Carlgren et al., 2020; Schindler et al., 2019). Blood-based assays for amyloid and tau are among the most exciting advances, offering minimally invasive and scalable tools for population-level screening. Table 1.3 states the AT(N) framework summarizing various markers (Brinkmalm et al., 2014; Kester et al., 2015).

Table 1.3 The AT(N) framework summarizing amyloid, tau, and neurodegeneration markers

Category	Modalities	Role in staging
A (Amyloid)	Amyloid PET, CSF Aβ42, plasma Aβ ratios	Earliest detectable pathology
T (Tau)	Tau PET, CSF p-Tau, plasma p-Tau	Correlates with symptom severity
N (Neurodegeneration)	MRI atrophy, FDG-PET, hypometabolism, CSF/plasma-NfL	Reflects downstream neuronal injury

1.4.1 Clinical Utility of Biomarkers

In the clinic, biomarkers serve several functions. They aid in differential diagnosis, distinguishing AD from other dementias such as frontotemporal dementia or vascular dementia (Preische et al., 2019). They provide prognostic information, identifying individuals at highest risk of progression. They are also essential for therapeutic monitoring, particularly in the era of disease-modifying therapies (DMTs) targeting amyloid and tau. Neuroimaging has revolutionized the study of AD progression, providing noninvasive windows into structural, functional, and molecular brain changes (Arbabshirani et al., 2017). Amyloid PET tracers such as florbetapir and Pittsburgh Compound B allow visualization of amyloid deposition in vivo (Mueller et al., 2005). Amyloid accumulation begins years before symptoms, plateauing by the time dementia is established. Thus, amyloid PET is valuable for early detection and differential diagnosis but less useful for tracking disease progression once dementia is present (Landau & Jagust, 2015). Tau PET imaging provides a more dynamic picture of disease progression. Tau deposition follows a Braak-like pattern, correlating closely with clinical severity and cognitive decline (Talbot et al., 2012). Tau PET is therefore emerging as a superior tool for staging and for monitoring progression in clinical trials (Ossenkoppele et al., 2016; Schöll et al., 2016). MRI remains a cornerstone in AD evaluation. Hippocampal and medial temporal lobe atrophy are classic findings. As disease advances, cortical thinning and ventricular enlargement become apparent. MRI also detects vascular lesions and other pathologies that may contribute to cognitive decline.

FDG-PET reveals hypometabolism in temporoparietal and posterior cingulate regions, often preceding structural atrophy (Launer et al., 2010). Functional MRI studies demonstrate disrupted connectivity within default mode and other networks, reflecting synaptic dysfunction (Jack et al., 2010). Combining imaging modalities offers a comprehensive perspective. For example, amyloid PET can identify early pathology, tau PET can stage progression, MRI can measure neurodegeneration, and FDG-PET can assess functional decline (Fan et al., 2015). Together, these tools form the backbone of modern in vivo staging.

1.4.2 Molecular Progression: Amyloid, Tau, Synaptic Dysfunction, and Neuroinflammation

AD progression reflects a cascade of molecular events, though the precise sequence remains debated (Bader et al., 2020). The amyloid cascade hypothesis posits that amyloid-β accumulation initiates AD, leading to tau pathology, synaptic dysfunction, and neuronal loss (Ciryam et al., 2015; Nixon, R. A., 2013). Genetic evidence supports this view, as mutations in amyloid precursor protein and presenilin genes cause familial AD with early amyloid deposition (Montagne et al., 2015). However, amyloid burden alone correlates poorly with clinical decline, suggesting it is necessary but not sufficient (Greicius et al., 2004; Landau & Jagust, 2015). Tau aggregation is more closely linked with neuronal injury and cognitive impairment. Pathological tau spreads in a prion-like manner along neural circuits, consistent with Braak staging (Clavaguera et al., 2009). This progression from entorhinal cortex to hippocampus to neocortex mirrors the clinical trajectory from memory deficits to global dementia (Landau & Jagust, 2015; Ossenkoppele et al., 2016; Villemagne et al., 2013). Synaptic loss is the best correlate of cognitive impairment in AD. Amyloid and tau pathology both contribute to synaptic dysfunction, disrupting neurotransmission and plasticity (Brinkmalm et al., 2014). Biomarkers of synaptic injury, such as neurogranin and synaptotagmin, are being developed to measure this critical process. Neuroinflammation plays a dual role. Activated microglia and astrocytes initially help clear amyloid but can become chronically activated, releasing pro-inflammatory cytokines and exacerbating neurodegeneration (Keren-Shaul et al., 2017). Genetic studies implicating microglial genes such as TREM2 underscore the importance of immune pathways in AD progression (Guerreiro et al., 2013; Heneka et al., 2015a; Karikari et al., 2020; Sweeney et al., 2018).

The progression of AD varies widely among individuals. Some decline rapidly, while others experience slower trajectories. Several factors influence this heterogeneity (Jack et al., 2019; Ngandu et al., 2015). The APOE ε4 allele is the strongest genetic risk factor for late-onset AD (Liu et al., 2013). Carriers develop amyloid earlier and decline more rapidly. Other genetic variants affecting inflammation, cholesterol metabolism, and synaptic function also modulate progression (Bellenguez et al., 2022; Kunkle et al., 2019). Hypertension, diabetes, and cerebrovascular disease accelerate cognitive decline (Valenzuela & Sachdev, 2006). Mixed pathologies are common in older adults and contribute to variability in clinical presentation (Iadecola et al., 2013). Younger onset cases often progress more rapidly. Women, particularly APOE ε4 carriers, may experience faster decline (Rosen et al., 1984). Cognitive reserve—shaped by education, occupation, and lifestyle—modifies resilience, with higher reserve delaying symptom onset despite equivalent pathology (Kourtis et al., 2019). Progression has traditionally been measured using cognitive and functional scales such as the Mini-Mental State Examination, Alzheimer's Disease Assessment Scale-Cognitive Subscale, and Clinical Dementia Rating (Blondell et al., 2014; Liang et al., 1986; Morris, J. C., 1993). These tools remain standard but have limitations, including floor and ceiling effects. With the rise of DMTs, biomarkers are increasingly used as endpoints in clinical trials. Amyloid PET, tau PET, and CSF measures provide objective evidence of target engagement and disease modification (Dunn et al., 2021). Blood biomarkers may soon serve as scalable alternatives (Hampel et al., 2018).

1.4.3 Challenges and Implications for Treatment and Clinical Trials

Challenges include variability in biomarker assays, the need for longitudinal validation, and the ethical implications of disclosing biomarker status to asymptomatic individuals (Milne et al., 2018). Despite these challenges, staging frameworks based on biomarkers are reshaping trial design and clinical practice. The recognition that AD begins long before dementia has shifted therapeutic strategies toward early intervention (Keren-Shaul et al., 2017). Anti-amyloid and anti-tau therapies are most likely to be effective when administered in preclinical or prodromal stages (Bereczki et al., 2018; Kuiper et al., 2015). Staging frameworks enable precise selection of participants who have biological evidence of AD but remain relatively preserved cognitively. In addition, staging helps monitor treatment response and safety (Ciryam et al., 2015). For example, amyloid-related imaging abnormalities are important side effects of anti-amyloid therapies, and biomarker monitoring is essential. Beyond DMTs, staging informs symptomatic management, caregiver planning, and palliative care (Andrieu et al., 2017). Table 1.4 reflects the stage-specific therapeutic targets in AD. Figures 1.4 and 1.5 depict the integration of clinical, neuropathological, and biomarker frameworks and the precision medicine pipeline integrating multimodal data for AD, respectively.

Table 1.4 Stage-specific therapeutic targets in AD

Stage	Preventive strategies	Disease-modifying	Symptomatic	Support
Preclinical	Lifestyle modification, risk reduction	Anti-amyloid, anti-tau	None	Education
MCI	Cognitive training	Early DMTs	Mild cognitive enhancers	Monitoring
Dementia	Supportive environment	Symptomatic + experimental	Cholinesterase inhibitors, memantine	Full caregiving

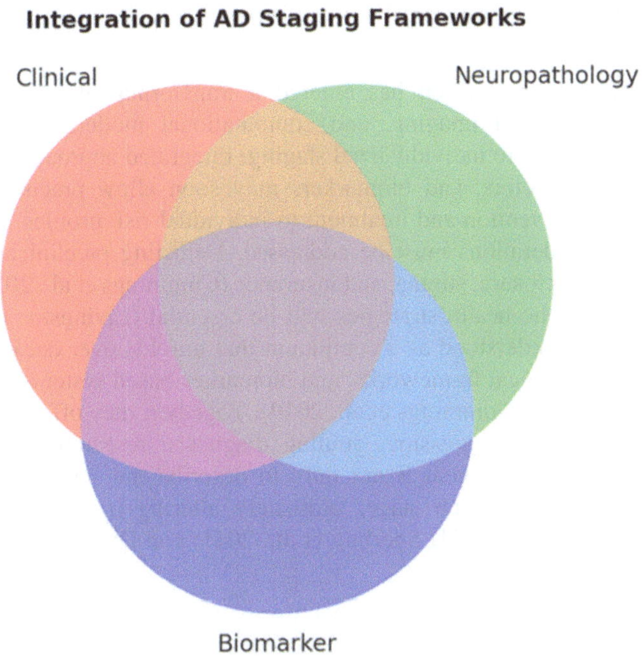

Fig. 1.4 Integration of clinical, neuropathological, and biomarker frameworks

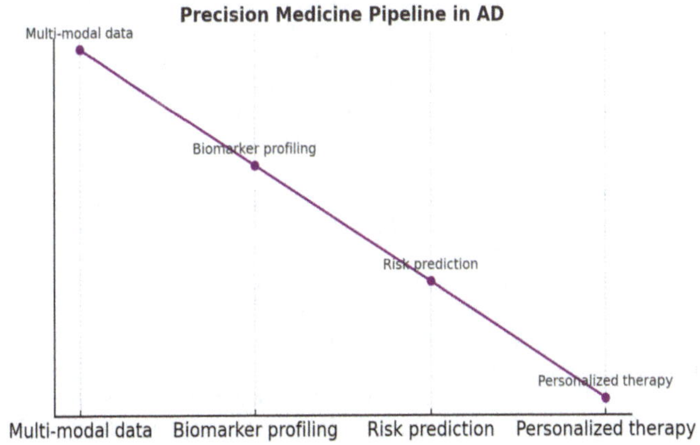

Fig. 1.5 Precision medicine pipeline integrating multimodal data for AD

1.5 Future Directions and Conclusion

The study of AD progression has entered a transformative era. Blood-based biomarkers, multimodal imaging, and computational models promise earlier, more accurate, and more individualized staging. Integrated approaches combining genetics, lifestyle factors, and biomarkers may soon allow precision medicine in AD, tailoring prevention and treatment to individual risk profiles. At the same time, ethical considerations must be addressed. Detecting preclinical AD raises questions about disclosure, stigma, and insurance (Cummings et al., 2022). Careful counseling and public health strategies will be essential (Livingston et al., 2017, 2020). AD is best understood as a continuum that unfolds over decades. Clinical stages, neuropathological frameworks, and biomarker-based systems each provide valuable perspectives (Cummings et al., 2019). Together, they offer a comprehensive view of disease progression, guiding diagnosis, research, and treatment. As science advances, the goal is not only to describe the stages of AD but to intervene effectively at each stage, ultimately altering its devastating course (Cummings et al., 2020, 2021; Sperling et al., 2011; van Dyck et al., 2023).

References

Albert, M. S., DeKosky, S. T., Dickson, D., Dubois, B., Feldman, H. H., Fox, N. C., Gamst, A., Holtzman, D. M., Jagust, W. J., Petersen, R. C., Snyder, P. J., Carrillo, M. C., Thies, B., & Phelps, C. H. (2011). The diagnosis of mild cognitive impairment due to Alzheimer's disease: Recommendations from the National Institute on Aging–Alzheimer's Association workgroups. *Alzheimer's & Dementia*, *7*(3), 270–279. https://doi.org/10.1016/j.jalz.2011.03.008

Andrieu, S., Guyonnet, S., Coley, N., Cantet, C., Bonnefoy, M., Bordes, S., Bories, L., Cufi, M. N., Dantoine, T., Dartigues, J. F., Desclaux, F., Gabelle, A., Gasnier, Y., Pesce, A., Sudres, K., Touchon, J., Robert, P., Rouaud, O., Legrand, P., & Vellas, B. (2017). Effect of long-term omega 3 polyunsaturated fatty acid supplementation with or without multidomain intervention on cognitive function in elderly adults. *JAMA*, *317*(6), 590–600. https://doi.org/10.1001/jama.2016.17277

Arbabshirani, M. R., Plis, S., Sui, J., & Calhoun, V. D. (2017). Single subject prediction of brain disorders in neuroimaging: Promises and pitfalls. *NeuroImage*, *145*, 137–165. https://doi.org/10.1016/j.neuroimage.2016.02.079

Arvanitakis, Z., Shah, R. C., & Bennett, D. A. (2019). Diagnosis and management of dementia: Review. *JAMA*, *322*(16), 1589–1599. https://doi.org/10.1001/jama.2019.4782

Ashley, E. A. (2016). Towards precision medicine. *Nature Reviews Genetics*, *17*(9), 507–522. https://doi.org/10.1038/nrg.2016.86

Bader, J. M., Geyer, P. E., Müller, J. B., Strauss, M. T., Koch, M., Leypoldt, F., Koertvelyessy, P., Bittner, D., Schipke, C. G., Incesoy, E. I., Peters, O., Deigendesch, N., Arlt, S., Janitzky, K., Mertens, A., Wurster, U., Walther, P., Albrecht, P., Mann, M., & Meissner, F. (2020). Proteome profiling in cerebrospinal fluid reveals novel biomarkers of Alzheimer's disease. *Molecular Systems Biology*, *16*(6), e9356. https://doi.org/10.15252/msb.20199356

Barabási, A. L., Gulbahce, N., & Loscalzo, J. (2011). Network medicine: A network-based approach to human disease. *Nature Reviews Genetics*, *12*(1), 56–68. https://doi.org/10.1038/nrg2918

Bellenguez, C., Küçükali, F., Jansen, I. E., Kleineidam, L., Moreno-Grau, S., Amin, N., Naj, A. C., Grenier-Boley, B., Andrade, V., Holmans, P., & Amouyel, P. (2022). New insights into the genetic etiology of Alzheimer's disease and related dementias. *Nature Genetics*, *54*(4), 412–436. https://doi.org/10.1038/s41588-022-01024-z

Bennett, D. A., Schneider, J. A., Arvanitakis, Z., & Wilson, R. S. (2012a). Overview and findings from the Religious Orders Study. *Current Alzheimer Research*, *9*(6), 628–645. https://doi.org/10.2174/156720512801322573

Bennett, D. A., Wilson, R. S., Boyle, P. A., Buchman, A. S., & Schneider, J. A. (2012b). Relation of neuropathology to cognition in persons without cognitive impairment. *Annals of Neurology*, *72*(4), 599–609. https://doi.org/10.1002/ana.23654

Bereczki, E., Branca, R. M., Francis, P. T., Pereira, J. B., Baek, J. H., Hortobágyi, T., Winblad, B., Ballard, C., Lehtiö, J., & Aarsland, D. (2018). Synaptic markers of cognitive decline in neurodegenerative diseases. *Nature Medicine*, *24*(4), 491–497. https://doi.org/10.1038/nm.4510

Blondell, S. J., Hammersley-Mather, R., & Veerman, J. L. (2014). Does physical activity prevent cognitive decline? *BMC Public Health*, *14*, 510. https://doi.org/10.1186/1471-2458-14-510

Braak, H., & Braak, E. (1991). Neuropathological stageing of Alzheimer-related changes. *Acta Neuropathologica*, *82*(4), 239–259. https://doi.org/10.1007/BF00308809

Brinkmalm, A., Brinkmalm, G., Honer, W. G., Frölich, L., Hausner, L., Minthon, L., Hansson, O., & Blennow, K. (2014). SNAP-25 is a promising novel cerebrospinal fluid biomarker for synapse degeneration in Alzheimer's disease. *Molecular Neurodegeneration*, *9*, 53. https://doi.org/10.1186/1750-1326-9-53

Brookmeyer, R., Corrada, M. M., Curriero, F. C., & Kawas, C. (2002). Survival following a diagnosis of Alzheimer disease. *Archives of Neurology*, *59*(11), 1764–1767. https://doi.org/10.1001/archneur.59.11.1764

Brookmeyer, R., Johnson, E., Ziegler-Graham, K., & Arrighi, H. M. (2007). Forecasting the global burden of Alzheimer's disease. *Alzheimer's & Dementia*, *3*(3), 186–191. https://doi.org/10.1016/j.jalz.2007.04.381

Butterfield, D. A., & Halliwell, B. (2019). Oxidative stress, dysfunctional glucose metabolism and Alzheimer disease. *Nature Reviews Neuroscience*, *20*(3), 148–160. https://doi.org/10.1038/s41583-019-0132-6

Cheng, G., Huang, C., Deng, H., & Wang, H. (2012). Diabetes as a risk factor for dementia and mild cognitive impairment: A meta-analysis of longitudinal studies. *Internal Medicine Journal*, *42*(5), 484–491. https://doi.org/10.1111/j.1445-5994.2012.02755.x

Ciryam, P., Kundra, R., Morimoto, R. I., Dobson, C. M., & Vendruscolo, M. (2015). Supersaturation is a major driving force for protein aggregation in neurodegenerative diseases. *Trends in Pharmacological Sciences*, *36*(2), 72–77. https://doi.org/10.1016/j.tips.2014.12.004

Clavaguera, F., Bolmont, T., Crowther, R. A., Abramowski, D., Frank, S., Probst, A., Fraser, G., Stalder, A. K., Beibel, M., Staufenbiel, M., Jucker, M., & Goedert, M. (2009). Transmission and spreading of tauopathy in transgenic mouse brain. *Nature Cell Biology*, *11*(7), 909–913. https://doi.org/10.1038/ncb1901

Cummings, J. (2022). Aducanumab: Appropriate use recommendations. *Journal of Prevention of Alzheimer's Disease*, *9*(2), 221–230. https://doi.org/10.14283/jpad.2022.18

Cummings, J., Aisen, P., Lemere, C., Atri, A., Sabbagh, M., & Salloway, S. (2020). Aducanumab produced a clinically meaningful benefit in association with amyloid lowering. *Alzheimer's Research & Therapy*, *12*(1), 98. https://doi.org/10.1186/s13195-020-00662-4

Cummings, J., Feldman, H. H., & Scheltens, P. (2019). The rights of precision drug development for Alzheimer's disease. *Alzheimer's Research & Therapy*, *11*, 76. https://doi.org/10.1186/s13195-019-0529-5

Cummings, J., Lee, G., Zhong, K., Fonseca, J., & Taghva, K. (2021). Alzheimer's disease drug development pipeline: 2021. *Alzheimer's & Dementia: Translational Research & Clinical Interventions*, *7*(1), e12179. https://doi.org/10.1002/trc2.12179

Crutch, S. J., Schott, J. M., Rabinovici, G. D., Boeve, B. F., Cappa, S. F., Dickerson, B. C., Dubois, B., Graff-Radford, J., Krolak-Salmon, P., Lehmann, M., Mendez, M. F., Pijnenburg, Y., Ryan, N. S., Scheltens, P., Shakespeare, T. J., Suárez González, A., Tang-Wai, D. F., van der Flier, W. M., & Fox, N. C. (2017). Consensus classification of posterior cortical atrophy. *Lancet Neurology*, *16*(2), 170–178. https://doi.org/10.1016/S1474-4422(16)30354-3

Donohue, M. C., Sperling, R. A., Salmon, D. P., Rentz, D. M., Raman, R., Thomas, R. G., Weiner, M., Aisen, P. S., & Alzheimer's Disease Cooperative Study. (2014). The preclinical Alzheimer cognitive composite. *Alzheimer's & Dementia*, *10*(6), 637–649. https://doi.org/10.1016/j.jalz.2014.02.014

Dunn, B., Stein, P., & Cavazzoni, P. (2021). Approval of aducanumab for Alzheimer disease—The FDA's perspective. *New England Journal of Medicine*, *385*(9), 771–774. https://doi.org/10.1056/NEJMp2109570

Fan, Z., Okello, A., Brooks, D. J., & Edison, P. (2015). An early and late peak in microglial activation in Alzheimer's disease trajectory. *Brain*, *138*(3), 822–834. https://doi.org/10.1093/brain/awu349

Folstein, M. F., Folstein, S. E., & McHugh, P. R. (1975). Mini-mental state: A practical method for grading the cognitive state of patients. *Journal of Psychiatric Research*, *12*(3), 189–198. https://doi.org/10.1016/0022-3956(75)90026-6

Greicius, M. D., Srivastava, G., Reiss, A. L., & Menon, V. (2004). Default-mode network activity distinguishes Alzheimer's disease from healthy aging. *Proceedings of the National Academy of Sciences*, *101*(13), 4637–4642. https://doi.org/10.1073/pnas.0308627101

Guerreiro, R., Wojtas, A., Bras, J., Carrasquillo, M., Rogaeva, E., Majounie, E., Cruchaga, C., Sassi, C., Kauwe, J. S. K., Younkin, S., & Hardy, J. (2013). TREM2 variants in Alzheimer's disease. *New England Journal of Medicine*, *368*(2), 117–127. https://doi.org/10.1056/NEJMoa1211851

Guo, J. L., & Lee, V. M. (2014). Cell-to-cell transmission of pathogenic proteins in neurodegenerative diseases. *Nature Medicine*, *20*(2), 130–138. https://doi.org/10.1038/nm.3457

Hampel, H., O'Bryant, S. E., Molinuevo, J. L., Zetterberg, H., Masters, C. L., Lista, S., Kiddle, S. J., Batrla, R., Blennow, K., & Biomarker Working Group. (2018). Blood-based biomarkers for Alzheimer disease: Mapping the road to the clinic. *Nature Reviews Neurology*, *14*(11), 639–652. https://doi.org/10.1038/s41582-018-0079-7

Hampel, H., Vergallo, A., Perry, G., & Lista, S. (2019). The Alzheimer precision medicine initiative. *Journal of Alzheimer's Disease*, *68*(1), 1–24. https://doi.org/10.3233/JAD-181121

Hansson, O., Zetterberg, H., Buchhave, P., Londos, E., Blennow, K., & Minthon, L. (2006). Association between CSF biomarkers and incipient Alzheimer's disease. *Lancet Neurology*, *5*(3), 228–234. https://doi.org/10.1016/S1474-4422(06)70355-6

Heneka, M. T., Carson, M. J., El Khoury, J., Landreth, G. E., Brosseron, F., Feinstein, D. L., Jacobs, A. H., Wyss-Coray, T., Vitorica, J., Ransohoff, R. M., Herrup, K., Frautschy, S. A., Finsen, B., Brown, G. C., Verkhratsky, A., Yamanaka, K., Koistinaho, J., Latz, E., Halle, A., & Kummer, M. P. (2015a). Neuroinflammation in Alzheimer's disease. *Lancet Neurology*, *14*(4), 388–405. https://doi.org/10.1016/S1474-4422(15)70016-5

Heneka, M. T., Golenbock, D. T., & Latz, E. (2015b). Innate immunity in Alzheimer's disease. *Nature Immunology*, *16*(3), 229–236. https://doi.org/10.1038/ni.3102

Iadecola, C. (2013). The pathobiology of vascular dementia. *Neuron*, *80*(4), 844–866. https://doi.org/10.1016/j.neuron.2013.10.008

Irwin, D. J., Lee, V. M., & Trojanowski, J. Q. (2013). Parkinson's disease dementia: Convergence of α-synuclein, tau and amyloid-β pathologies. *Nature Reviews Neuroscience*, *14*(9), 626–636. https://doi.org/10.1038/nrn3549

Jack, C. R., Bennett, D. A., Blennow, K., Carrillo, M. C., Dunn, B., Haeberlein, S. B., Holtzman, D. M., Jagust, W., Jessen, F., Karlawish, J., Liu, E., Molinuevo, J. L., Montine, T., Phelps, C., Rankin, K. P., Rowe, C. C., Scheltens, P., Siemers, E., Snyder, H. M., & Sperling, R. (2018). NIA-AA research framework: Toward a biological definition of Alzheimer's disease. *Alzheimer's & Dementia*, *14*(4), 535–562. https://doi.org/10.1016/j.jalz.2018.02.018

Jack, C. R., Knopman, D. S., Jagust, W. J., Shaw, L. M., Aisen, P. S., Weiner, M. W., Petersen, R. C., & Trojanowski, J. Q. (2010). Hypothetical model of dynamic biomarkers of the Alzheimer's pathological cascade. *Lancet Neurology*, *9*(1), 119–128. https://doi.org/10.1016/S1474-4422(09)70299-6

Jack, C. R., Therneau, T. M., Wiste, H. J., Weigand, S. D., Knopman, D. S., Vemuri, P., Mielke, M. M., Lowe, V. J., Senjem, M. L., Gunter, J. L., Machulda, M. M., Roberts, R. O., Rocca, W. A., & Petersen, R. C. (2016). Associations of amyloid, tau, and neurodegeneration biomarker profiles with rates of memory decline among individuals without dementia. *JAMA*, *315*(18), 2017–2025. https://doi.org/10.1001/jama.2016.4668

Jack, C. R., Wiste, H. J., Therneau, T. M., Weigand, S. D., Knopman, D. S., Mielke, M. M., Lowe, V. J., Vemuri, P., Machulda, M. M., Schwarz, C. G., Senjem, M. L., Gunter, J. L., Rocca, W. A., Petersen, R. C., & Roberts, R. O. (2019). Associations of amyloid, tau, and neurodegeneration biomarker profiles with rates of memory decline among individuals without dementia. *JAMA*, *321*(23), 2316–2325. https://doi.org/10.1001/jama.2019.7437

Janelidze, S., Mattsson-Carlgren, N., Palmqvist, S., Smith, R., Beach, T. G., Serrano, G. E., Chai, X., Proctor, N. K., Eichenlaub, U., Zetterberg, H., Blennow, K., & Hansson, O. (2020). Plasma p-tau217 detects Alzheimer disease in early and preclinical stages. *Nature Medicine*, *26*(3), 379–386. https://doi.org/10.1038/s41591-020-0763-1

Jessen, F., Amariglio, R. E., van Boxtel, M., Breteler, M., Ceccaldi, M., Chételat, G., Dubois, B., Dufouil, C., Ellis, K. A., van der Flier, W. M., Glodzik, L., van Harten, A. C., de Leon, M. J., McHugh, P., Mielke, M. M., Molinuevo, J. L., Mosconi, L., Osorio, R. S., Perrotin, A., & Wagner, M. (2014). A conceptual framework for research on subjective cognitive decline in preclinical Alzheimer's disease. *Alzheimer's & Dementia*, *10*(6), 844–852. https://doi.org/10.1016/j.jalz.2014.01.001

Johnson, E. C. B., Dammer, E. B., Duong, D. M., Ping, L., Zhou, M., Yin, L., Higginbotham, L., Guajardo, A., White, B., Troncoso, J. C., Thambisetty, M., Montine, T. J., Lee, E. B., Trojanowski, J. Q., & Seyfried, N. T. (2020). Large-scale proteomic analysis of Alzheimer's disease brain and cerebrospinal fluid reveals early changes in energy metabolism associated with microglia and astrocyte activation. *Nature Medicine*, *26*(5), 769–780. https://doi.org/10.1038/s41591-020-0815-6

Johnson, K. A., Schultz, A., Betensky, R. A., Becker, J. A., Sepulcre, J., Rentz, D., Mormino, E., Chhatwal, J., Amariglio, R., Papp, K., Marshall, G., Albers, M., Mauro, S., Pepin, L., Alverio, J., Judge, K., Philiossaint, M., Shoup, T., Yokell, D., & Sperling, R. (2016). Tau positron emission tomographic imaging in aging and early Alzheimer disease. *Annals of Neurology*, *79*(1), 110–119. https://doi.org/10.1002/ana.24546

Kapasi, A., DeCarli, C., & Schneider, J. A. (2017). Impact of multiple pathologies on the threshold for clinically overt dementia. *Acta Neuropathologica*, *134*(2), 171–186. https://doi.org/10.1007/s00401-017-1717-7

Karikari, T. K., Pascoal, T. A., Ashton, N. J., Janelidze, S., Benedet, A. L., Rodriguez, J. L., Chamoun, M., Savard, M., Kang, M. S., Therriault, J., Schöll, M., Massarweh, G., Soucy, J. P., Höglund, K., Brinkmalm, G., Mattsson-Carlgren, N., Palmqvist, S., Blennow, K., Zetterberg, H., & Hansson, O. (2020). Blood phosphorylated tau 181 as a biomarker for Alzheimer's disease. *Nature Medicine*, *26*(3), 387–397. https://doi.org/10.1038/s41591-020-0755-1

Keren-Shaul, H., Spinrad, A., Weiner, A., Matcovitch-Natan, O., Dvir-Szternfeld, R., Ulland, T. K., David, E., Baruch, K., Lara-Astaiso, D., Toth, B., Itzkovitz, S., Colonna, M., Schwartz, M., & Amit, I. (2017). A unique microglia type associated with restricting development of Alzheimer's disease. *Cell*, *169*(7), 1276–1290. https://doi.org/10.1016/j.cell.2017.05.018

Kester, M. I., Teunissen, C. E., Crimmins, D. L., Herries, E. M., Ladenson, J. H., Scheltens, P., van der Flier, W. M., Morris, J. C., & Holtzman, D. M. (2015). Neurogranin as a cerebrospinal fluid biomarker for synaptic loss in symptomatic Alzheimer disease. *JAMA Neurology*, *72*(11), 1275–1280. https://doi.org/10.1001/jamaneurol.2015.1867

Klunk, W. E., Koeppe, R. A., Price, J. C., Benzinger, T. L., Devous, M. D., Jagust, W. J., Johnson, K. A., Mathis, C. A., Minhas, D., Pontecorvo, M. J., Rowe, C. C., Skovronsky, D. M., & Mintun, M. A. (2015). The Centiloid project: Standardizing quantitative amyloid plaque estimation by PET. *Alzheimer's & Dementia*, *11*(1), 1–15. https://doi.org/10.1016/j.jalz.2014.07.003

Kourtis, L. C., Regele, O. B., Wright, J. M., & Jones, G. B. (2019). Digital biomarkers for Alzheimer's disease. *NPJ Digital Medicine*, *2*, 9. https://doi.org/10.1038/s41746-019-0084-2

Kuiper, J. S., Zuidersma, M., Oude Voshaar, R. C., Zuidema, S. U., van den Heuvel, E. R., Stolk, R. P., & Smidt, N. (2015). Social relationships and risk of dementia: A systematic review and meta-analysis of longitudinal cohort studies. *Ageing Research Reviews*, *22*, 39–57. https://doi.org/10.1016/j.arr.2015.04.006

Kunkle, B. W., Grenier-Boley, B., Sims, R., Bis, J. C., Damotte, V., Naj, A. C., Boland, A., Vronskaya, M., van der Lee, S. J., Amlie-Wolf, A., & Pericak-Vance, M. A. (2019). Genetic meta-analysis of diagnosed Alzheimer's disease identifies new risk loci and implicates Aβ, tau, immunity and lipid processing. *Nature Genetics*, *51*(3), 414–430. https://doi.org/10.1038/s41588-019-0358-2

Landau, S. M., & Jagust, W. J. (2015). Imaging biomarkers of Alzheimer's disease and risk. *Lancet Neurology*, *14*(8), 782–794. https://doi.org/10.1016/S1474-4422(15)00080-7

Launer, L. J., Hughes, T., Yu, B., Masaki, K., Petrovitch, H., Ross, G. W., & White, L. R. (2010). Lowering midlife levels of systolic blood pressure as a public health strategy to reduce late-life dementia. *JAMA*, *304*(19), 2173–2179. https://doi.org/10.1001/jama.2010.1667

Lehmann, M., Crutch, S. J., Ridgway, G. R., Ridha, B. H., Barnes, J., Warrington, E. K., Rossor, M. N., & Fox, N. C. (2011). Cortical thickness and voxel-based morphometry in posterior cortical atrophy. *Brain*, *134*(5), 1468–1480. https://doi.org/10.1093/brain/awr040

Lewczuk, P., Lelental, N., Spitzer, P., Maler, J. M., & Kornhuber, J. (2015). Amyloid-β 42/40 cerebrospinal fluid concentration ratio in the diagnostics of Alzheimer's disease: Validation of two novel assays. *Journal of Alzheimer's Disease*, *43*(1), 183–191. https://doi.org/10.3233/JAD-140771

Liang, K. Y., & Zeger, S. L. (1986). Longitudinal data analysis using generalized linear models. *Biometrika*, *73*(1), 13–22. https://doi.org/10.1093/biomet/73.1.13

Lin, P. J., Zhong, Y., Fillit, H. M., Cohen, J. T., & Neumann, P. J. (2016). Medicare expenditures of individuals with Alzheimer's disease and related dementias or mild cognitive impairment before and after diagnosis. *Journal of the American Geriatrics Society*, *64*(8), 1549–1557. https://doi.org/10.1111/jgs.14227

Liu, C. C., Liu, C. C., Kanekiyo, T., Xu, H., & Bu, G. (2013). Apolipoprotein E and Alzheimer disease: Risk, mechanisms and therapy. *Nature Reviews Neurology*, *9*(2), 106–118. https://doi.org/10.1038/nrneurol.2012.263

Livingston, G., Huntley, J., Sommerlad, A., Ames, D., Ballard, C., Banerjee, S., Brayne, C., Burns, A., Cohen-Mansfield, J., Cooper, C., Costafreda, S. G., Dias, A., Fox, N., Gitlin, L. N., Howard, R., Kales, H. C., Kivimäki, M., Larson, E. B., Ogunniyi, A., & Mukadam, N. (2020). Dementia prevention, intervention, and care: 2020 report of the Lancet Commission. *The Lancet*, *396*(10248), 413–446. https://doi.org/10.1016/S0140-6736(20)30367-6

Livingston, G., Sommerlad, A., Orgeta, V., Costafreda, S. G., Huntley, J., Ames, D., Ballard, C., Banerjee, S., Burns, A., Cohen-Mansfield, J., Cooper, C., Fox, N., Gitlin, L. N., Howard, R., Kales, H. C., Larson, E. B., Ritchie, K., Rockwood, K., Sampson, E. L., & Mukadam, N. (2017). Dementia prevention, intervention, and care. *The Lancet*, *390*(10113), 2673–2734. https://doi.org/10.1016/S0140-6736(17)31363-6

Lyketsos, C. G., Carrillo, M. C., Ryan, J. M., Khachaturian, A. S., Trzepacz, P., Amatniek, J., Cedarbaum, J., Brashear, R., & Miller, D. S. (2011). Neuropsychiatric symptoms in Alzheimer's disease. *Alzheimer's & Dementia*, *7*(5), 532–539. https://doi.org/10.1016/j.jalz.2011.05.2410

Mathys, H., Davila-Velderrain, J., Peng, Z., Gao, F., Mohammadi, S., Young, J. Z., Menon, M., He, L., Abdurrob, F., Jiang, X., Martorell, A. J., Ransohoff, R. M., Hafler, B. P., Bennett, D. A., Kellis, M., & Tsai, L. H. (2019). Single-cell transcriptomic analysis of Alzheimer's disease. *Nature*, *570*(7761), 332–337. https://doi.org/10.1038/s41586-019-1195-2

Mattsson, N., Cullen, N. C., Andreasson, U., Zetterberg, H., & Blennow, K. (2019). Association between longitudinal plasma neurofilament light and neurodegeneration in patients with Alzheimer disease. *JAMA Neurology*, *76*(7), 791–799. https://doi.org/10.1001/jamaneurol.2019.0765

Mattsson-Carlgren, N., Palmqvist, S., Janelidze, S., & Hansson, O. (2020). Longitudinal plasma p-tau217 is increased in early stages of Alzheimer's disease. *Brain*, 143(11), 3234–3241. https://doi.org/10.1093/brain/awaa286

Mesulam, M. M., Weintraub, S., Rogalski, E. J., Wieneke, C., Geula, C., & Bigio, E. H. (2014). Asymmetry and heterogeneity of Alzheimer's and frontotemporal pathology in primary progressive aphasia. *Brain*, *137*(4), 1176–1192. https://doi.org/10.1093/brain/awu024

Milne, R., Bunnik, E., Diaz, A., Richard, E., & Badger, S. (2018). Ethical issues in the use of biomarkers for Alzheimer's disease. *Alzheimer's & Dementia*, *14*(7), 1107–1115. https://doi.org/10.1016/j.jalz.2018.04.003

Mintun, M. A., Lo, A. C., Duggan Evans, C., Wessels, A. M., Ardayfio, P. A., Andersen, S. W., Shcherbinin, S., Sparks, J., Sims, J. R., Brys, M., Apostolova, L. G., Salloway, S., Skovronsky, D. M., & TRAILBLAZER-ALZ 2 Investigators. (2021). Donanemab in early Alzheimer's disease. *New England Journal of Medicine*, *384*(18), 1691–1704. https://doi.org/10.1056/NEJMoa2100708

Mitchell, A. J., & Shiri-Feshki, M. (2009). Rate of progression of mild cognitive impairment to dementia. *Acta Psychiatrica Scandinavica*, *119*(4), 252–265. https://doi.org/10.1111/j.1600-0447.2008.01326.x

Montagne, A., Nation, D. A., Pa, J., Sweeney, M. D., Toga, A. W., & Zlokovic, B. V. (2015). Brain imaging of neurovascular dysfunction in Alzheimer's disease. *Neuron*, *85*(2), 296–302. https://doi.org/10.1016/j.neuron.2014.12.041

Morris, J. C. (1993). The Clinical Dementia Rating (CDR): Current version and scoring rules. *Neurology*, *43*(11), 2412–2414. https://doi.org/10.1212/WNL.43.11.2412-a

Mueller, S. G., Weiner, M. W., Thal, L. J., Petersen, R. C., Jack, C., Jagust, W., Trojanowski, J. Q., Toga, A. W., & Beckett, L. (2005). The Alzheimer's Disease Neuroimaging Initiative. *Neuroimaging Clinics of North America*, *15*(4), 869–877. https://doi.org/10.1016/j.nic.2005.09.008

Ngandu, T., Lehtisalo, J., Solomon, A., Levälahti, E., Ahtiluoto, S., Antikainen, R., Bäckman, L., Hänninen, T., Jula, A., Laatikainen, T., Lindström, J., Mangialasche, F., Paajanen, T., Pajala, S., Peltonen, M., Rauramaa, R., Stigsdotter-Neely, A., Strandberg, T., Tuomilehto, J., & Kivipelto, M. (2015). A 2-year multidomain intervention of diet, exercise, cognitive training, and vascular risk monitoring versus control. *The Lancet*, *385*(9984), 2255–2263. https://doi.org/10.1016/S0140-6736(15)60461-5

Nixon, R. A. (2013). The role of autophagy in neurodegenerative disease. Nature Medicine, *19*(8), 983–997. https://doi.org/10.1038/nm.3232

Ossenkoppele, R., Schonhaut, D. R., Schöll, M., Lockhart, S. N., Ayakta, N., Baker, S. L., O'Neil, J. P., Janabi, M., Lazaris, A., Cantwell, A., Vogel, J., Santos, M., Miller, Z., Rosen, H., Vossel, K., Miller, B. L., Jagust, W. J., & Rabinovici, G. D. (2016). Tau PET patterns mirror clinical and neuroanatomical variability in Alzheimer's disease. *Brain*, *139*(5), 1551–1567. https://doi.org/10.1093/brain/aww027

Palmqvist, S., Janelidze, S., Quiroz, Y. T., Zetterberg, H., Lopera, F., Stomrud, E., Su, Y., Chen, Y., Serrano, G. E., Leuzy, A., Mattsson-Carlgren, N., Strandberg, O., Smith, R., Villegas, A., Sepulveda-Falla, D., Chai, X., Proctor, N. K., Beach, T. G., Blennow, K., & Hansson, O. (2020). Discriminative accuracy of plasma phosphotau217 for Alzheimer disease vs other neurodegenerative disorders. *JAMA*, *324*(8), 772–781. https://doi.org/10.1001/jama.2020.12134

Petersen, R. C., Smith, G. E., Waring, S. C., Ivnik, R. J., Tangalos, E. G., & Kokmen, E. (1999). Mild cognitive impairment: Clinical characterization and outcome. *Archives of Neurology*, *56* (3), 303–308. https://doi.org/10.1001/archneur.56.3.303

Preische, O., Schultz, S. A., Apel, A., Kuhle, J., Kaeser, S. A., Barro, C., Gräber, S., Kuder-Buletta, E., LaFougere, C., Laske, C., Vöglein, J., Levin, J., Duering, M., Liesz, A., Giese, A., Peters, O., Härtl, E., Rami, A., Zetterberg, H., & Blennow, K. (2019). Serum neurofilament dynamics predicts neurodegeneration in presymptomatic Alzheimer's disease. *Nature Medicine*, *25*(2), 277–283. https://doi.org/10.1038/s41591-018-0304-3

Roberts, J. S., & Connell, C. M. (2000). Illness representations among first-degree relatives of people with Alzheimer disease. *Alzheimer Disease & Associated Disorders*, *14*(3), 129–136. https://doi.org/10.1097/00002093-200007000-00003

Rosen, W. G., Mohs, R. C., & Davis, K. L. (1984). A new rating scale for Alzheimer's disease. *American Journal of Psychiatry*, *141*(11), 1356–1364. https://doi.org/10.1176/ajp.141.11.1356

Sachdev, P., Kalaria, R., O'Brien, J., Skoog, I., Alladi, S., Black, S. E., Blacker, D., Blazer, D. G., Chen, C., Chui, H., Ganguli, M., Jellinger, K., Jorm, A., Korczyn, A., Passmore, P., & Rockwood, K. (2014). Diagnostic criteria for vascular cognitive disorders. *Alzheimer's & Dementia*, *10*(3), 243–261. https://doi.org/10.1016/j.jalz.2014.01.007

Schindler, S. E., Bollinger, J. G., Ovod, V., Mawuenyega, K. G., Li, Y., Gordon, B. A., Hassenstab, J., Xiong, C., Fagan, A. M., & Bateman, R. J. (2019). High-precision plasma β-amyloid 42/40 predicts current and future brain amyloidosis. *Neurology*, *93*(17), e1647–e1659. https://doi.org/10.1212/WNL.0000000000008081

Schöll, M., Lockhart, S. N., Schonhaut, D. R., O'Neil, J. P., Janabi, M., Ossenkoppele, R., Baker, S. L., Vogel, J. W., Faria, J., Schwimmer, H. D., Rabinovici, G. D., & Jagust, W. J. (2016). PET imaging of tau deposition in the aging human brain. *Neuron*, *89*(5), 971–982. https://doi.org/10.1016/j.neuron.2016.01.028

Sperling, R. A., Aisen, P. S., Beckett, L. A., Bennett, D. A., Craft, S., Fagan, A. M., Iwatsubo, T., Jack, C. R., Kaye, J., Montine, T. J., Park, D. C., Reiman, E. M., Rowe, C. C., Siemers, E., Stern, Y., & Yaffe, K. (2011). Toward defining the preclinical stages of Alzheimer's disease. *Alzheimer's & Dementia*, *7*(3), 280–292. https://doi.org/10.1016/j.jalz.2011.03.003

Sweeney, M. D., Sagare, A. P., & Zlokovic, B. V. (2018). Blood-brain barrier breakdown in Alzheimer disease and other neurodegenerative disorders. *Nature Reviews Neurology*, *14*(3), 133–150. https://doi.org/10.1038/nrneurol.2017.188

Talbot, K., Wang, H. Y., Kazi, H., Han, L. Y., Bakshi, K. P., Stucky, A., Fuino, R. L., Kawaguchi, K. R., Samoyedny, A. J., Wilson, R. S., Arvanitakis, Z., Schneider, J. A., Wolf, B. A., Bennett, D. A., Trojanowski, J. Q., & Arnold, S. E. (2012). Demonstrated brain insulin resistance in Alzheimer's disease. *Journal of Clinical Investigation*, 122(4), 1316–1338. https://doi.org/10.1172/JCI59903

Thal, D. R., Rüb, U., Orantes, M., & Braak, H. (2002). Phases of Aβ-deposition in the human brain and its relevance for the development of Alzheimer's disease. *Neurology*, *58*(12), 1791–1800. https://doi.org/10.1212/WNL.58.12.1791

van Dyck, C. H., Swanson, C. J., Aisen, P., Bateman, R. J., Chen, C., Gee, M., Kanekiyo, M., Li, D., Reyderman, L., Cohen, S., Froelich, L., Katayama, S., Sabbagh, M., Vellas, B., Watson, D., Dhadda, S., Irizarry, M., Kramer, L. D., & Iwatsubo, T. (2023). Lecanemab in early Alzheimer's disease. *New England Journal of Medicine*, 388(1), 9–21. https://doi.org/10.1056/NEJMoa2212948

Valenzuela, M. J., & Sachdev, P. (2006). Brain reserve and dementia: A systematic review. *Psychological Medicine*, *36*(4), 441–454. https://doi.org/10.1017/S0033291705006264

Villemagne, V. L., Burnham, S., Bourgeat, P., Brown, B., Ellis, K. A., Salvado, O., Szoeke, C., Macaulay, S. L., Martins, R., Maruff, P., Ames, D., Rowe, C. C., & Masters, C. L. (2013). Amyloid β deposition, neurodegeneration, and cognitive decline in sporadic Alzheimer's disease. *Lancet Neurology*, *12*(4), 357–367. https://doi.org/10.1016/S1474-4422(13)70044-9

Zetterberg, H., & Blennow, K. (2021). From cerebrospinal fluid to blood: The third wave of fluid biomarkers for Alzheimer's disease. *Nature Reviews Neurology*, 17(11), 703–714. https://doi.org/10.1038/s41582-021-00549-x

Chapter 2
Alzheimer's Disease-Associated Cognitive Disabilities

Subhayan Das and Prashant Mishra

Abstract Alzheimer's disease is a type of neurodegenerative disorder that usually occurs in older age. Although the hallmark of the disease is memory impairment, it is often associated with other cognitive and behavioral impairments. The disease most commonly presents as insidiously progressing memory loss that slowly progresses toward other impairments like language disorders (anomia or aphasia), attention deficits, and other cognitive impairments. As the disease progresses, the patients often experience extrapyramidal dysfunctions like dystonia, bradykinesia, tremor, and even Parkinsonism. Usually, in the early phases of the disease, social and behavioral impairments are not significant. However, as the disease progresses, the patient's ability to participate in day-to-day activities is impaired, leading to substantial limitations in social and occupational activities. These impairments include confusion about location, taking longer to complete daily activities, difficulties in handling money, mood swings, and personality changes. In the late stage of the disease, the patient cannot read or write correctly, has problems recognizing friends and family members, and has difficulties in logical thinking and reasoning. In extreme cases, the patient often lacks bowel and bladder control, loss of motor functions, and extrapyramidal symptoms.

S. Das (✉) · P. Mishra
Department of Allied Health Sciences, Brainware University, Barasat, Kolkata, West Bengal, India
e-mail: subhayan81@gmail.com

M. Kundu, S. Das (eds.), *Alzheimer's Disease-Associated Cognitive Disabilities*, SpringerBriefs in Modern Perspectives on Disability Research,
https://doi.org/10.1007/978-981-95-7121-5_2

Graphical Abstract

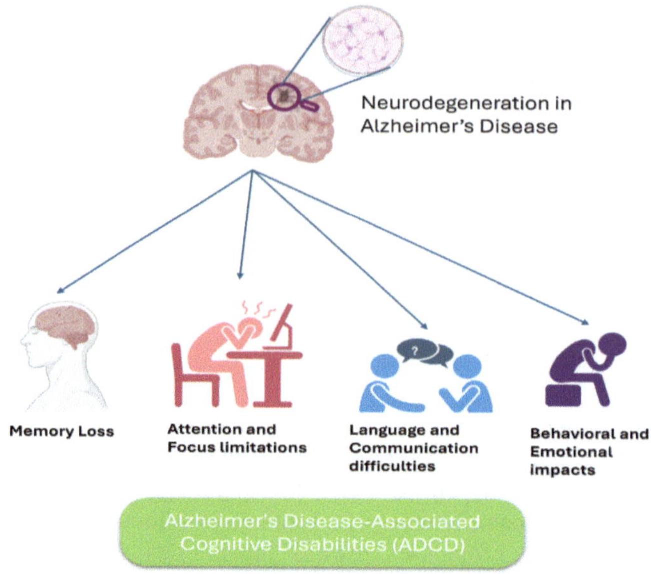

Keywords Cognitive disabilities · Memory loss · Language deficits · Alzheimer's associated behavioral disorders · Speech disorders

2.1 Introduction

Cognitive disabilities are conditions that affect higher brain functions like thinking, learning, memory, and problem-solving. Cognitive disabilities are commonly seen in diseases like autism, attention-deficit/hyperactivity disorder (ADHD), traumatic brain injury (TBI), and dementia. Genetic factors, trauma, and neurodegenerative disorders are main contributors to cognitive disorders (Levine et al., 2023). Alzheimer's disease (AD) is one of the most prominent causes of dementia globally. It is a progressive neurodegenerative disorder affecting nearly 30 million people worldwide. Its pathogenesis shows loss of synapses and neurons (Kamatham et al., 2024). The younger age groups show a very low or minimal rate of AD, while the disease chance significantly jumps from 5% to 50% from age 70 to 90 ("2024 Alzheimer's Disease Facts and Figures," 2024).

One of the clinical hallmarks of AD is the extracellular plaque formation and neurofibrillary tangles inside the microtubules of the neurons which firstly accumulate in the entorhinal cortex and hippocampal formations before spreading to temporal, parietal, and frontal association cortices. Highly myelinated neurons remain relatively spared until late stages, whereas poorly myelinated limbic neurons—critical for memory and learning—are affected early (Kuznetsov & Kuznetsov, 2018). Subcortical neuron loss in the nucleus basalis of Meynert and locus coeruleus further impairs cholinergic and noradrenergic transmission. In healthy aging, annual brain volume loss averages 0.2–0.41% (Lagarde et al., 2024).

Since AD is a neurodegenerative disorder, it is frequently associated with different cognitive impairments involving memory, logic, and behavioral changes. Memory loss, challenges with reasoning and problem-solving, and difficulties with language and judgment are among the progressive cognitive changes associated with AD (Tenchov et al., 2024). These changes are brought on by the death of brain cells. Among these impairments, the inability to remember new information is frequently the initial symptom that worsens over time, interfering with day-to-day activities and ultimately leading to serious impairments in executive functions and the capacity to carry out daily tasks (Safiri et al., 2024). The progression of cognitive impairments is seen in AD in three stages. In the early stage, the changes are subtle and confined to memory loss and complex tasks. In the middle stages, the changes are more prominent, affecting multiple aspects of memory loss and sleep disorders, along with behavioral and emotional changes. In the late stage, the impairments are so severe that day-to-day activities are hampered. There are severe memory loss and cognitive disabilities that may lead to inability to identify common objects, recognize close relatives, or perform basic tasks (Ávila-Villanueva et al., 2022). Cognitive disabilities associated with stages of AD are summarized in Table 2.1.

2.2 Memory Loss and Recall Impairments

Memory loss is one of the most prominent features of cognitive impairment in AD. Memory impairment can be seen primarily in long-term and short-term memory impairments. Additionally, impairment of sensory memory like memory of smell and sleep memory can also be impaired.

2.2.1 Short-Term Memory

An important early sign of AD is short-term memory loss, which is defined by the inability to recall recent discussions, occasions, or recently acquired knowledge. AD-related memory loss is chronic and worsens over time, impairing a person's capacity to carry out daily tasks (Silva et al., 2019). Finding the appropriate words, getting confused in familiar settings, and having trouble with complex tasks are some of the other symptoms that frequently accompany this (Yeung, 2024). Short-term memory loss usually appears early in the course of AD preceding the long-term memory loss. In later stages, it can impair higher functions like judgment, planning, and logical thinking. In this stage, it may become difficult to perform certain tasks like cooking, driving, or managing the finances (Soncu Büyükişcan, 2025).

Table 2.1 Overview of stage-based impairments in AD

Stage	Key cognitive & behavioral characteristics	Key pathological & functional changes	References
Early stage	1. Subtle memory loss (especially short-term) 2. Difficulty with complex tasks (finances, planning) 3. Difficulties in word-finding (anomia) 4. Confusion in familiar settings	1. Initial accumulation of amyloid plaques & neurofibrillary tangles in entorhinal cortex hippocampus 2. Social/behavioral impairments are not significant	Yeung, 2024; Silva et al., 2019
Middle stage	1. More prominent memory loss (affecting long-term memory) 2. Language disorders (aphasia) 3. Attention deficits 4. Sleep disturbances and mood swings 5. Personality changes 6. Apathy 7. Anxiety	1. Pathology spreads to temporal, parietal, and frontal cortices 2. Disruption of the default mode network (DMN) becomes evident 3. Significant impairment in daily activities	Ávila-Villanueva et al., 2022; Mortby et al., 2022; Moore & Henneghan, 2023
Late stage	1. Severe memory & cognitive disability 2. Inability to recognize objects/relatives 3. Loss of logical thinking, loss of speech 4. Psychotic symptoms (delusions, hallucinations) 5. Agitation 6. Aggression	1. Widespread cortical atrophy and extrapyramidal symptoms (tremor, bradykinesia) 2. Loss of bowel/bladder control 3. Patient is fully dependent on care	Ismail et al., 2022; Burhanullah et al., 2020

2.2.2 Long-Term Memory

Long-term memory loss is also a characteristic feature of AD that usually manifests in the advanced stage of the disease. Long-term memory loss often includes forgetting past events, places, or even relationships. It often includes inability to recall childhood events, family history, or significant life events. This stage may also include forgetting names and places including direction or orientation (Ma et al., 2023). There is significant social impairment resulting from long-term memory loss in AD. This may start with misplacing objects or forgetting names of family members. However, in advanced stages, a person can forget the direction and become confused and disoriented significantly impeding daily activity (Tahami Monfared et al., 2022).

2.2.3 *Memory of Smell*

A meta-analysis suggests that dysfunction of the olfactory can be preclinical sign of AD. Early pathology in the entorhinal cortex is responsible for processing smell. The meta-analysis concluded that AD patients show more deficits in smelling scents and flavors linked to smell which impairs the ability of odor recognition and identification (Silva et al., 2018).

Twenty percent of the AD patients are expected to have a seizure, and in some genetic variability (presenilin-1 mutations or trisomy 21), there is a chance of unprovoked seizure up to 84%. It is more common during the early onset of AD, and these nonconclusive epileptiform discharges in many cases have led to cognitive impairment or disturbances causing episodes of amnestic wandering (Yang et al., 2021).

Increased beta-amyloid activity leads to neuronal excitability which activates the strengthening of the inhibitory networks which leads to new synaptic organization. This cascade is believed to activate the cognitive decline and compromise the functioning of the DMN (Musaeus et al., 2024).

2.2.4 *Sleep and Memory*

Clinically, sleep–wake disturbances—like excessive daytime napping and insomnia—affect 25–40% of mild-to-moderate AD patients. Even in amnestic mild cognitive impairment, altered sleep architecture and EEG patterns are present and may worsen memory by disrupting sleep-dependent consolidation (Chen et al., 2024).

In a study of 145 cognitively healthy adults over 45, 32 individuals with amyloid deposition showed poorer sleep efficiency, despite similar total sleep times, and frequent napping (three or more times per week) was linked to amyloid presence. These findings suggest that patterns of brain activity during sleep can influence molecular processes underlying disease (Ju et al., 2013). In AD, heightened resting-state activity may promote amyloid buildup, pointing to potential interventions that target brain metabolism or emphasize healthy sleep habits for prevention.

2.2.5 *Default Mode Network and its Role in Alzheimer's Disease*

The DMN is a set of brain regions that remain active during resting states and internally oriented cognitive processes. It encompasses structures such as the medial temporal lobe, medial prefrontal cortex, posterior cingulate cortex, ventral precuneus, and both medial and lateral parietal cortices. In AD, converging evidence demonstrates that DMN undergoes progressive disruption affecting both structure and function.

Metabolic imaging with fluorodeoxyglucose positron emission tomography (FDG-PET) consistently identifies reduced glucose uptake in posterior DMN nodes, particularly the posterior cingulate, retrosplenial cortex, inferior parietal lobule, and lateral temporal cortex—changes that parallel the course of cognitive decline. Individuals carrying the ApoE4 allele often show early metabolic disturbances and diminished resting-state connectivity within the DMN, sometimes preceding detectable amyloid plaque accumulation. Functional MRI further illustrates abnormal DMN behavior, with reduced intrinsic synchrony and impaired task-related deactivation, pointing to instability in normal network dynamics.

Neuropathological findings also support the network's selective vulnerability. Amyloid deposits accumulate preferentially within DMN hubs, suggesting that the network's high synaptic activity and metabolic demands may drive amyloid genesis and accelerate damage within memory circuits. Structural imaging adds a complementary perspective, revealing accelerated atrophy in the medial temporal lobe and posterior cingulate during preclinical stages. Together, these alterations provide a mechanistic link between DMN breakdown and the early memory deficits that characterize AD progression.

2.3 Language and Communication Difficulties

Language and speech impairments are one of the key features of AD with symptoms that can be seen in the early phases which will see a gradual decline as the disease progresses and language and speech impairments worsen over the latter stages of AD (Volkmer et al., 2024).

2.3.1 *Progressive Language Deterioration Across AD Stages*

The impairments are prevalent during the early stages primarily affecting the socio-linguistic skills where patients experience confusion with word meanings, naming, and identifying objects and face difficulties with word comprehension (Klimova et al., 2015). Studies showed that mild AD patients maintained smooth conversation flow and showed preserved vocabulary skills for known concepts during early phases, but AD patients in later stages could not maintain the same flow and used meaningless words which hinted toward the decline of communication skills as it became fragmented, and the word flow was hindered (Vigo et al., 2022). In some other studies, it was found that complex language processing across all stages was preserved with only some issues on unknown linguistic expression (Chapin et al., 2022).

Different types of language difficulties in AD include anomia, aphasia, echolalia, decrease in fluency, and nonverbal communication (Vigo et al., 2022). In anomia, the person is unable to name an object or find the right word, often replacing it with an inappropriate substitute. In aphasia, the person is unable to express something in words and is often unable to speak. In echolalia, a person repeats a particular phrase or words without making any sense. Most of these impairments are found in advanced stages of AD increasing the difficulty in social and emotional interactions (Cruz, 2010; Silagi et al., 2015; Wilson et al., 2018).

2.3.2 Quantitative Assessment of Language Decline

A cross-sectional study of 35 AD patients (20 mild vs 15 moderate) has shown a systematic pattern of impairments upon oral, comprehensive, and language areas. In this study, semantic word processing was seen to decline the most (69.46 ± 4.48 points in mild AD to 55.82 ± 13.93 points in moderate AD ($p = 0.001$). This high difference highlights the severity of the issue of compromised ability of understanding meaning of words in moderate AD patients. Other test tasks that showed a decline were the visual confrontation naming task which showed a significant decline (105.0 ± 9.26 vs 85.75 ± 15.54, $p < 0.001$) with patients having severe anomia and experiencing verbal paraphasia. The critical working memory impairment was interpreted as the complex ideational material comprehension also showed a homologous decline (7.78 ± 1.47 vs 6.26 ± 1.24 points, $p = 0.003$) (Ortiz et al., 2024).

2.3.3 Machine Learning Analysis of Speech Features

Machine learning techniques and advanced computational analysis methods were applied to study speech impairments in AD, where 83 AD patients were studied vs 75 healthy controls using 3-minute speech samples, and the results revealed significant alterations in speech features. One of the key findings in spectral analysis was the significant difference in center of gravity (AD: 1230.9 Hz $\pm$ 182.7 vs healthy: 1546.1 Hz $\pm$ 211.5, $p < 0.001$), indicating impaired articulatory strength and precision, and the other key finding in this study was that the intensity measurements revealed compromised vocal control, particularly highlighting decreased minimum intensity in AD patients (0.002 vs 0.003, $p < 0.001$) which suggests vocal memory fades and breath support is compromised. The findings hint that AD affects speech production via cortical atrophy and neuro-impairments affecting motor planning and execution (Noto et al., 2024).

With the findings of these studies and many more reported symptoms across the globe, it is known that the language impairments in AD can occur at the earliest stages due to its underlying working memory dysfunction which underlies various language impairments making it difficult for AD patients to establish good communication and progress to lead a normal life.

2.4 Attention and Focus Limitations

The overtime accumulation of amyloid-beta plaques which worsens over time and the tangles in the microtubules of neurons, also known as neurofibrillary tangles, have a negative impact on the neuronal networks that are important for attention- and focus-related functioning. The locus coeruleus is essential for regulating the noradrenergic pathway which signals for arousal, and alerting is one of the first brain regions to be affected by tau pathophysiology; additionally, the basal forebrain cholinergic system can also be seen to degenerate which further compromises the attentional neuronal network. The heavy disruption of both the ascending stem pathways and cortical connections between the frontal and parietal regions due to the pathophysiological degeneration leads to numerous attention- and focus-related impairments.

2.4.1 *Selective Attention Deficits*

In the cases of patients with AD, the selective attention impairment generally targeted the ability of the patient to focus on the main thing or the relevant stimuli. The patients were unable to filter out the distraction stimuli and maintain focus on the main specific thing (Chau et al., 2015). Studies demonstrate that while patients could utilize the color cues to maintain focus attention on relevant array components, they could not ignore irrelevant stimuli in complex visual research tasks (Eraslan Boz et al., 2024; Polden et al., 2020). Another example of this would be an impaired AD patient not being able to maintain conversation in a noisy area as they will not be able to focus on the external noises (Banovic et al., 2018).

2.4.2 *Sustained Attention Dysfunction*

There is significant sustained memory or vigilance in the case of AD patients compared against the healthy controls. Studies showed that mild AD patients have shown increased sensitivity decrement over time while performing high tasks that demand a lot of higher processing, especially in conditions or environments that require greater effortful processing (Liu et al., 2024). This greatly affects how long a patient is able to maintain focus during prolonged attention-demanding activities and has shown to have a negative impact. The sustained attention deficits are seen to appear earlier than any other attention deficits with their symptoms being prevalent as early as the pre-onset or preclinical phase of AD (Siéroff et al., 2020).

2.4.3 Divided Attention Impairments

The most severely impaired domain in the case of AD patients is the divided attention when patients could not focus on two or more things at the same time. Research has shown that the patients faced difficulties when required to simultaneously respond to multiple stimuli and perform various tasks at a multitasking level. From the neuroimaging studies, we can conclude that cortical activation had become weaker, and it required activation of more brain regions compared to healthy controls while performing divided attention tasks (Rad et al., 2025).

2.4.4 Executive Attention and Orienting Deficits

The executive attention networks which include conflict resolution and attentional controls are mainly compromised through anterior cingulate and prefrontal dysfunction. The orienting network responsible for spatial attention is mainly affected in posterior cortical atrophy variants of AD, where patients manifest features of Balint's syndrome where patients show symptoms of simultanagnosia and attentional disengagement difficulties. These deficits greatly contribute to the visuospatial impairments which reach far beyond visual modalities to auditory spatial processing (Kurth et al., 2019; Ng et al., 2021).

Alzheimer's disease attentional deficits are the result of a complex interaction between several different neural networks and pathological alterations. The importance of comprehending and treating attentional dysfunction in comprehensive Alzheimer's care is highlighted by the fact that these deficiencies not only exacerbate cognitive decline but also have a substantial impact on patients' capacity to carry out activities of daily living and preserve their independence.

2.5 Behavioral and Emotional Impacts

2.5.1 *Depression and Anxiety*

Depression is observed in roughly 15% of individuals with AD when focusing on major depressive episodes, and the rate increases further when milder cases are considered. In dementia, however, depressive features are often present in atypical ways—patients seldom verbalize feelings of sadness (McDade et al., 2020). Instead, they more commonly display loss of interest or pleasure, heightened preoccupation with physical complaints, and symptoms linked to anxiety. Anxiety itself is highly prevalent in AD, ranking as the third most frequent behavioral disturbance, with estimates suggesting it affects about 40% of patients (Moore & Henneghan, 2023).

2.5.2 *Apathy*

Recognized as the most common behavioral disorder in AD, apathy affects up to 90% of patients with severe dementia. This syndrome involves reduced motivation for goal-directed behaviors and cognitive activities, alongside blunted emotional responses. Apathy demonstrates the strongest correlation with cognitive decline among all behavioral symptoms, predicting progression from normal cognition to mild cognitive impairment and subsequently to dementia (Mortby et al., 2022).

2.5.3 *Emotional Lability and Reactivity*

Patients frequently experience rapid emotional shifts and mood lability characterized by sudden changes within seconds or minutes. Studies suggest preserved emotional reactivity to stimuli, particularly positive emotions, though emotional stimuli often fail to capture sustained attention despite evidence of initial emotional response (Chaudhary et al., 2022).

AD patients also show some psychotic behavioral patterns which encompass around 41% of AD patients. These symptoms are delusions which can be in the form of theft and misidentification or even harassment of other people, and in some cases, hallucinations can also be seen. At the beginning, it is mainly visual but can gradually develop to more than that like auditory, smell, or other olfactory senses (Ismail et al., 2022).

2.5.4 *Psychotic and Behavioral Symptoms*

Psychosis: Psychotic symptoms occur in approximately 41% of AD patients, with delusions affecting 36% and hallucinations 18%. Delusions typically involve themes of theft, abandonment, or misidentification, while hallucinations are predominantly visual but can encompass auditory, olfactory, or tactile modalities (Ismail et al., 2022).

Table 2.2 Summary of key cognitive and behavioral impairments in AD

Domain	Specific impairments	Key findings/Manifestations	References
Memory	Short-term memory loss	Inability to recall recent events/ conversations; appears early and precedes long-term memory loss	Silva et al., 2019; Yeung, 2024
	Long-term memory loss	Forgetting past events, relationships, and personal history; leads to significant social impairment and disorientation	Ma et al., 2023; Tahami Monfared et al., 2022
	Memory of smell	Olfactory dysfunction as a preclinical sign; impaired odor recognition and identification	Silva et al., 2018
Language	Anomia & aphasia	Inability to name objects or find words; inability to express oneself; worsens with disease progression	Volkmer et al., 2024; Cruz, 2010
	Quantitative decline	Significant decline in semantic word processing and visual confrontation naming from mild-to-moderate AD	Ortiz et al., 2024
	Speech production	Altered speech features (e.g., lower center of gravity, decreased intensity) due to cortical atrophy and motor planning deficits	Noto et al., 2024
Attention	Selective attention	Inability to filter out distractions and focus on relevant stimuli	Chau et al., 2015; Eraslan Boz et al., 2024
	Sustained attention	Difficulty maintaining focus over time during prolonged tasks; appears very early in the disease	Liu et al., 2024; Siéroff et al., 2020
	Divided attention	Severe inability to multitask or respond to multiple stimuli simultaneously	Rad et al., 2025
Behavior and emotion	Apathy	Most common behavioral disorder; reduced motivation and blunted emotional response; strongly correlated with cognitive decline	Mortby et al., 2022
	Depression & anxiety	Prevalent, often presenting atypically with loss of interest and physical complaints	McDade et al., 2020; Moore & Henneghan, 2023
	Psychosis & agitation	Delusions (theft, misidentification) and hallucinations (primarily visual); agitation includes restlessness and aggression	Ismail et al., 2022; Burhanullah et al., 2020

Agitation and Aggression: Agitation represents a complex behavioral syndrome affecting over 50% of patients with mild AD, increasing with disease progression. The International Psychogeriatric Association defines agitation as inappropriate verbal, vocal, or motor activity encompassing restlessness, verbal outbursts, and physical aggression (Burhanullah et al., 2020).

2.6 Conclusion

Alzheimer's disease presents an array of progressive cognitive deficits impacting beyond the most recognized clinical symptom of memory loss. In this chapter, we have explained how the continuing accumulation of pathological proteins disrupts neural networks, with an initial involvement of the DMN. Key cognitive disorders associated with AD are summarized in Table 2.2. The interference in the DMN leads to a progression of interrelated impairments of the cognition, including memory encoding and retrieval, language, and attention control. As dementia progresses, cognitive impairment can occur, in addition to clinically significant behavioral and emotional disorders (e.g., apathy, depression, and psychosis) that significantly increase the risk of stress to caregivers and reduce quality of life for patients and families. The cumulative impact of dementia-associated neurodegeneration, network de-differentiation, and resultant cognitive and behavioral symptoms ultimately deprives individuals of independent function, illustrating the immense impact of AD on the lives of patients and underlining the importance of multidomain care and intervention strategies.

Acknowledgment The figures in this chapter are created in https://BioRender.com. We are grateful for their generous help.

References

2024 Alzheimer's disease facts and figures. (2024). *Alzheimer's & Dementia: The Journal of the Alzheimer's Association*, *20*(5), 3708–3821. https://doi.org/10.1002/alz.13809

Ávila-Villanueva, M., Marcos Dolado, A., Gómez-Ramírez, J., & Fernández-Blázquez, M. (2022). Brain structural and functional changes in cognitive impairment due to Alzheimer's disease. *Frontiers in Psychology*, *13*, https://doi.org/10.3389/fpsyg.2022.886619

Banovic, S., Zunic, L., & Sinanovic, O. (2018). Communication difficulties as a result of dementia. *Materia Socio Medica*, *30*(2), 221. https://doi.org/10.5455/msm.2018.30.221-224

Burhanullah, M. H., Tschanz, J. T., Peters, M. E., Leoutsakos, J.-M., Matyi, J., Lyketsos, C. G., Nowrangi, M. A., & Rosenberg, P. B. (2020). Neuropsychiatric symptoms as risk factors for cognitive decline in clinically normal older adults: The Cache County Study. *The American Journal of Geriatric Psychiatry*, *28*(1), 64–71. https://doi.org/10.1016/j.jagp.2019.03.023

Chapin, K., Clarke, N., Garrard, P., & Hinzen, W. (2022). A finer-grained linguistic profile of Alzheimer's disease and mild cognitive impairment. *Journal of Neurolinguistics*, *63*, 101069. https://doi.org/10.1016/j.jneuroling.2022.101069

Chau, S. A., Herrmann, N., Eizenman, M., Chung, J., & Lanctôt, K. L. (2015). Exploring visual selective attention towards novel stimuli in Alzheimer's disease patients. *Dementia and Geriatric Cognitive Disorders Extra*, *5*(3), 492–502. https://doi.org/10.1159/000442383

Chaudhary, S., Zhornitsky, S., Chao, H. H., van Dyck, C. H., & Li, C.-S. R. (2022). Emotion processing dysfunction in Alzheimer's disease: An overview of behavioral findings, systems neural correlates, and underlying neural biology. *American Journal of Alzheimer's Disease & Other Dementias®*, *37*. https://doi.org/10.1177/15333175221082834

Chen, J., Peng, G., & Sun, B. (2024). Alzheimer's disease and sleep disorders: A bidirectional relationship. *Neuroscience*, *557*, 12–23. https://doi.org/10.1016/j.neuroscience.2024.08.008

Cruz, F. M. D. (2010). Verbal repetitions and echolalia in Alzheimer's discourse. *Clinical Linguistics & Phonetics*, *24*(11), 848–858. https://doi.org/10.3109/02699206.2010.511403

Eraslan Boz, H., Koçoğlu, K., Akkoyun, M., Tüfekci, I. Y., Ekin, M., & Akdal, G. (2024). Visual search in Alzheimer's disease and amnestic mild cognitive impairment: An eye-tracking study. *Alzheimer's & Dementia*, *20*(2), 759–768. https://doi.org/10.1002/alz.13478

Ismail, Z., Creese, B., Aarsland, D., Kales, H. C., Lyketsos, C. G., Sweet, R. A., & Ballard, C. (2022). Psychosis in Alzheimer disease — Mechanisms, genetics and therapeutic opportunities. *Nature Reviews Neurology*, *18*(3), 131–144. https://doi.org/10.1038/s41582-021-00597-3

Ju, Y.-E. S., McLeland, J. S., Toedebusch, C. D., Xiong, C., Fagan, A. M., Duntley, S. P., Morris, J. C., & Holtzman, D. M. (2013). Sleep Quality and Preclinical Alzheimer Disease. *JAMA Neurology*, *70*(5), 587. https://doi.org/10.1001/jamaneurol.2013.2334

Kamatham, P. T., Shukla, R., Khatri, D. K., & Vora, L. K. (2024). Pathogenesis, diagnostics, and therapeutics for Alzheimer's disease: Breaking the memory barrier. *Ageing Research Reviews*, *101*, 102481. https://doi.org/10.1016/j.arr.2024.102481

Klimova, B., Maresova, P., Valis, M., Hort, J., & Kuca, K. (2015). Alzheimer's disease and language impairments: Social intervention and medical treatment. *Clinical Interventions in Aging*, *10*, 1401–1407. https://doi.org/10.2147/CIA.S89714

Kurth, S., Bahri, M. A., Collette, F., Phillips, C., Majerus, S., Bastin, C., & Salmon, E. (2019). Alzheimer's disease patients activate attention networks in a short-term memory task. *NeuroImage: Clinical*, *23*, 101892. https://doi.org/10.1016/j.nicl.2019.101892

Kuznetsov, I. A., & Kuznetsov, A. V. (2018). How the formation of amyloid plaques and neurofibrillary tangles may be related: A mathematical modelling study. *Proceedings of the Royal Society A: Mathematical, Physical and Engineering Sciences*, *474*(2210), 20170777. https://doi.org/10.1098/rspa.2017.0777

Lagarde, J., Olivieri, P., Tonietto, M., Noiray, C., Lehericy, S., Valabrègue, R., Caillé, F., Gervais, P., Moussion, M., Bottlaender, M., & Sarazin, M. (2024). Combined in vivo MRI assessment of locus coeruleus and nucleus basalis of Meynert integrity in amnestic Alzheimer's disease, suspected-LATE and frontotemporal dementia. *Alzheimer's Research & Therapy*, *16*(1), 97. https://doi.org/10.1186/s13195-024-01466-z

Levine, S. Z., Rotstein, A., Kodesh, A., Sandin, S., Lee, B. K., Weinstein, G., Schnaider Beeri, M., & Reichenberg, A. (2023). Adult Attention-Deficit/Hyperactivity Disorder and the Risk of Dementia. *JAMA Network Open*, *6*(10), e2338088. https://doi.org/10.1001/jamanetworkopen.2023.38088

Liu, E., Zhang, Y., & Wang, J.-Z. (2024). Updates in Alzheimer's disease: From basic research to diagnosis and therapies. *Translational Neurodegeneration*, *13*(1), 45. https://doi.org/10.1186/s40035-024-00432-x

Ma, J., Zheng, M.-X., Wu, -J.-J., Xing, -X.-X., Xiang, Y.-T., Wei, D., Xue, X., Zhang, H., Hua, X.-Y., Guo, Q.-H., & Xu, J.-G. (2023). Mapping the long-term delayed recall-based cortex-hippocampus network constrained by the structural and functional connectome: A case-control multimodal MRI study. *Alzheimer's Research & Therapy*, *15*(1), 61. https://doi.org/10.1186/s13195-023-01197-7

McDade, E., Bednar, M. M., Brashear, H. R., Miller, D. S., Maruff, P., Randolph, C., Ismail, Z., Carrillo, M. C., Weber, C. J., Bain, L. J., & Hake, A. M. (2020). The pathway to secondary prevention of Alzheimer's disease. *Alzheimer's & Dementia: Translational Research & Clinical Interventions*, *6*(1). https://doi.org/10.1002/trc2.12069

Moore, R. C., & Henneghan, A. M. (2023). Examining individual differences in the within-person process of perceived stress and cognitive functioning to advance precision health: Commentary on "sex differences in the relationship between perceived stress and cognitive trajectories" by Paolillo et al. *The American Journal of Geriatric Psychiatry*, *31*(6), 411–414. https://doi.org/10.1016/j.jagp.2023.01.018

Mortby, M. E., Adler, L., Agüera-Ortiz, L., Bateman, D. R., Brodaty, H., Cantillon, M., Geda, Y. E., Ismail, Z., Lanctôt, K. L., Marshall, G. A., Padala, P. R., Politis, A., Rosenberg, P. B., Siarkos, K., Sultzer, D. L., & Theleritis, C. (2022). Apathy as a treatment target in Alzheimer's disease: Implications for clinical trials. *The American Journal of Geriatric Psychiatry*, *30*(2), 119–147. https://doi.org/10.1016/j.jagp.2021.06.016

Musaeus, C. S., Kjaer, T. W., Lindberg, U., Vestergaard, M. B., Bo, H., Larsson, W., Press, D. Z., Andersen, B. B., Høgh, P., Kidmose, P., Hemmsen, M. C., Rank, M. L., Hasselbalch, S. G., Waldemar, G., & Frederiksen, K. S. (2024). Subclinical epileptiform discharges in Alzheimer's disease are associated with increased hippocampal blood flow. *Alzheimer's Research & Therapy*, *16*(1), 80. https://doi.org/10.1186/s13195-024-01432-9

Ng, A. S. L., Wang, J., Ng, K. K., Chong, J. S. X., Qian, X., Lim, J. K. W., Tan, Y. J., Yong, A. C. W., Chander, R. J., Hameed, S., Ting, S. K. S., Kandiah, N., & Zhou, J. H. (2021). Distinct network topology in Alzheimer's disease and behavioral variant frontotemporal dementia. *Alzheimer's Research & Therapy*, *13*(1), 13. https://doi.org/10.1186/s13195-020-00752-w

Noto, S., Sekiyama, Y., Nagata, R., Yamamoto, G., & Tamura, T. (2024). Analysis of speech features in Alzheimer's disease with machine learning: A case-control study. *Healthcare*, *12*(21), 2194. https://doi.org/10.3390/healthcare12212194

Ortiz, K. Z., De Lira, J. O., Minett, T. S. C., & Bertolucci, P. H. F. (2024). Language impairments in Alzheimer´s disease: What changes can be found between mild and moderate stages of the disease? *Clinics*, *79*, 100412. https://doi.org/10.1016/j.clinsp.2024.100412

Polden, M., Wilcockson, T. D. W., & Crawford, T. J. (2020). The disengagement of visual attention: An eye-tracking study of cognitive impairment, ethnicity and age. *Brain Sciences*, *10*(7), 461. https://doi.org/10.3390/brainsci10070461

Rad, E. M., Mazinani, S. M., & Zendehbad, S. A. (2025). Diagnosis of Alzheimer's disease using non-linear features of ERP signals through a hybrid attention-based CNN-LSTM model. *Computer Methods and Programs in Biomedicine Update*, *7*, 100192. https://doi.org/10.1016/j.cmpbup.2025.100192

Safiri, S., Ghaffari Jolfayi, A., Fazlollahi, A., Morsali, S., Sarkesh, A., Daei Sorkhabi, A., Golabi, B., Aletaha, R., Motlagh Asghari, K., Hamidi, S., Mousavi, S. E., Jamalkhani, S., Karamzad, N., Shamekh, A., Mohammadinasab, R., Sullman, M. J. M., Şahin, F., & Kolahi, -A.-A. (2024). Alzheimer's disease: A comprehensive review of epidemiology, risk factors, symptoms diagnosis, management, caregiving, advanced treatments and associated challenges. *Frontiers in Medicine*, *11*. https://doi.org/10.3389/fmed.2024.1474043

Siéroff, É., Joly, C., & Lenoir, H. (2020). Cognitive aspects of attention deficits in Alzheimer's disease. *Gériatrie et Psychologie Neuropsychiatrie du Viellissement*, *18*(4), 419–428. https://doi.org/10.1684/pnv.2020.0898

Silagi, M. L., Bertolucci, P. H. F., & Ortiz, K. Z. (2015). Naming ability in patients with mild to moderate Alzheimer's disease: What changes occur with the evolution of the disease? *Clinics*, *70*(6), 423–428. https://doi.org/10.6061/clinics/2015(06)07

Silva, M. D. M. E., Mercer, P. B. S., Witt, M. C. Z., & Pessoa, R. R. (2018). Olfactory dysfunction in Alzheimer's disease Systematic review and meta-analysis. *Dementia & Neuropsychologia*, *12*(2), 123–132. https://doi.org/10.1590/1980-57642018dn12-020004

Silva, M. V. F., Loures, C. D. M. G., Alves, L. C. V., de Souza, L. C., Borges, K. B. G., & Carvalho, M. D. G. (2019). Alzheimer's disease: Risk factors and potentially protective measures. *Journal of Biomedical Science*, *26*(1), 33. https://doi.org/10.1186/s12929-019-0524-y

Soncu Büyükişcan, E. (2025). Neuropsychology of Alzheimer's disease: From preclinical phase to dementia. *Applied Neuropsychology: Adult*, 1–9. https://doi.org/10.1080/23279095.2025.2469236

Tahami Monfared, A. A., Byrnes, M. J., White, L. A., & Zhang, Q. (2022). The humanistic and economic burden of Alzheimer's disease. *Neurology and Therapy*, *11*(2), 525–551. https://doi.org/10.1007/s40120-022-00335-x

Tenchov, R., Sasso, J. M., & Zhou, Q. A. (2024). Alzheimer's Disease: Exploring the Landscape of Cognitive Decline. *ACS Chemical Neuroscience*, *15*(21), 3800–3827. https://doi.org/10.1021/acschemneuro.4c00339

Vigo, I., Coelho, L., & Reis, S. (2022). Speech- and Language-Based Classification of Alzheimer's Disease: A Systematic Review. *Bioengineering (Basel, Switzerland)*, *9*(1). https://doi.org/10.3390/bioengineering9010027

Volkmer, A., Cross, L., Highton, L., Jackson, C., Smith, C., Brotherhood, E., Harding, E. V., Mummery, C., Rohrer, J., Weil, R., Yong, K., Crutch, S., & Hardy, C. J. D. (2024). 'Communication is difficult': Speech, language and communication needs of people with young onset or rarer forms of non-language led dementia. *International Journal of Language & Communication Disorders*, *59*(4), 1553–1577. https://doi.org/10.1111/1460-6984.13018

Wilson, S. M., Yen, M., & Eriksson, D. K. (2018). An adaptive semantic matching paradigm for reliable and valid language mapping in individuals with aphasia. *Human Brain Mapping*, *39*(8), 3285–3307. https://doi.org/10.1002/hbm.24077

Yang, W., Li, X., Pan, K., Yang, R., Song, R., Qi, X., Pedersen, N. L., & Xu, W. (2021). Association of life-course depression with the risk of dementia in late life: A nationwide twin study. *Alzheimer's & Dementia*, *17*(8), 1383–1390. https://doi.org/10.1002/alz.12303

Yeung, M. K. (2024). Metamemory and executive function mediate the age-related decline in memory. *Journal of the International Neuropsychological Society*, *30*(5), 479–488. https://doi.org/10.1017/S1355617723011451

Chapter 3
Management of Alzheimer's Disease-Associated Cognitive Disabilities

Shiuli Bera and Moumita Kundu

Abstract Alzheimer's disease-associated cognitive disabilities pose an increasing challenge for healthcare systems worldwide due to their progressive nature and the significant burden they place on patients and their families. Effective management requires more than symptom control; it demands a comprehensive approach that integrates medical, psychological, and social strategies. Early and accurate diagnosis remains critical, with clinical evaluations, cognitive tests, neuroimaging, and biomarker studies providing the basis for timely and targeted intervention. Pharmacological treatments, such as cholinesterase inhibitors, can provide modest improvements in cognition and daily functioning, while emerging therapies directed at amyloid and tau pathology offer hope for modifying the underlying disease process. Equally important are non-pharmacological strategies. Cognitive rehabilitation programs, regular physical exercise, and thoughtful environmental adjustments help preserve independence and slow functional decline. Caregivers play an indispensable role in this process, as their well-being and preparedness directly influence patient outcomes. Counseling, structured support services, and practical training can ease the emotional and physical demands of long-term care. Taken together, these multidimensional strategies not only address clinical needs but also promote a better quality of life for individuals living with Alzheimer's disease-associated cognitive disabilities and those who care for them.

S. Bera
Department of Pharmaceutical Technology, Brainware University,
Barasat, Kolkata, West Bengal, India

M. Kundu (✉)
Department of Pharmaceutical Technology, Brainware University,
Barasat, Kolkata, West Bengal, India;
Center for Multidisciplinary Research and Innovations, Brainware University,
Barasat, Kolkata, West Bengal, India
e-mail: moumitakundu1991@gmail.com

M. Kundu, S. Das (eds.), *Alzheimer's Disease-Associated Cognitive Disabilities*,
SpringerBriefs in Modern Perspectives on Disability Research,
https://doi.org/10.1007/978-981-95-7121-5_3

Graphical Abstract

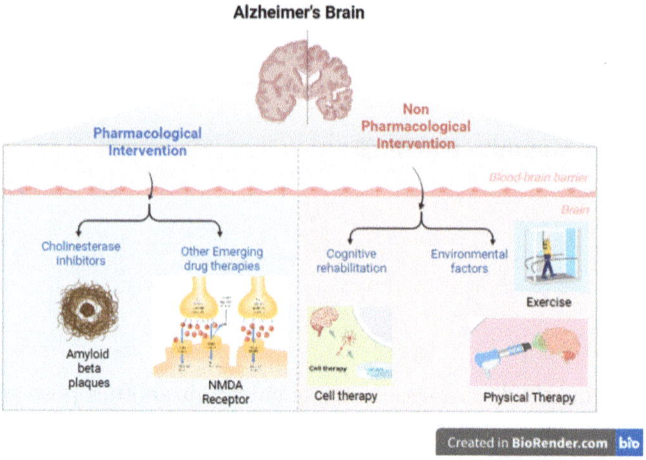

Keywords Alzheimer's disease · Cognitive impairment · Biomarkers · Pharmacological therapy · Non-pharmacological management

3.1 Introduction to Alzheimer's Disease-Associated Cognitive Disabilities Management

Alzheimer's disease (AD) is a progressive neurodegenerative disorder characterized primarily by the accumulation of amyloid-beta (Aβ) plaques and neurofibrillary tangles (NFTs) in the brain, which lead to synaptic dysfunction and cognitive decline (Lashley et al., 2018; Shah-Abadi et al., 2023) (Fig. 3.1). Although many studies have been conducted, establishing a definitive diagnosis of AD is still difficult to achieve, mainly due to inconsistencies between molecular findings and clinical presentation. At the same time, biomarkers have emerged as valuable tools for improving diagnostic accuracy. Among them, cerebrospinal fluid (CSF) biomarkers are regarded as particularly informative, as neuronal changes such as degeneration and protein alterations are often mirrored quickly in the CSF (Moghekar et al., 2013). AD unfolds as a chronic and progressive condition, gradually affecting multiple aspects of a person's life. Aside from its effects on daily functioning and behavior, it is primarily marked by a gradual decline in cognition, memory, and the emergence of neuropsychiatric symptoms (Scheltens et al., 2021). The 2018 World Alzheimer's Disease Report estimated that around 50 million people worldwide are living with dementia, with AD accounting for roughly two-thirds of these cases. In China alone, more than 6 million individuals are affected, and projections suggest that this number could surpass 10 million by 2050 (AsdtjotAs, 2018; Abbott, 2023). As one of the leading causes of disability

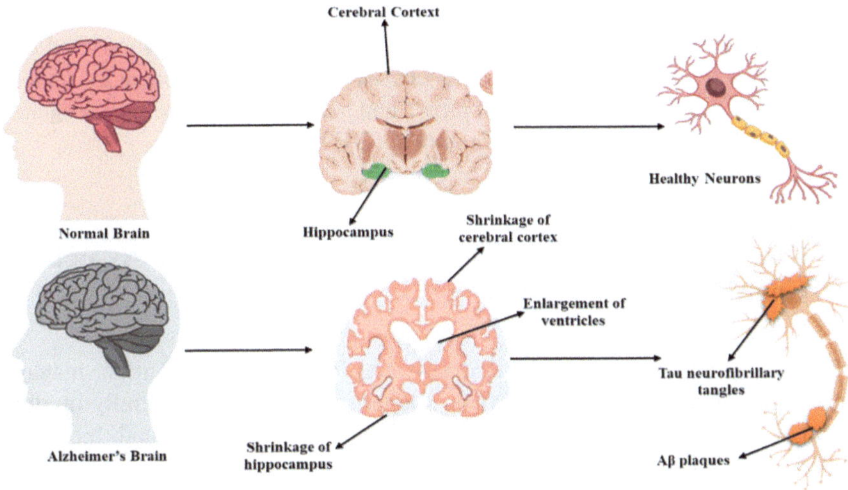

Fig. 3.1 Structural and functional brain changes in AD

and mortality among the elderly, AD poses enormous social and economic challenges, not only for patients but also for families, caregivers, healthcare systems, and society at large.

Because the disease begins subtly, progresses slowly, and arises from multiple causes, no treatment that can fully cure the condition is currently available (Wang et al., 2022). Management approaches are generally divided into two major categories: pharmacological treatments and supportive, non-pharmacological interventions, with medications remaining the central focus of medical management (Mangialasche et al., 2010). Commonly prescribed drugs include cholinesterase inhibitors (ChEIs), such as rivastigmine and galantamine, both approved by the U.S. Food and Drug Administration (FDA) as first-line therapies (Giacobini et al., 2022). Based on the amyloid-β (Aβ) hypothesis—which proposes that Aβ deposition induces tau hyperphosphorylation, leading to neuronal degeneration and cognitive impairment (Kamble et al., 2024)—monoclonal antibody therapies have been designed to target Aβ. Lecanemab and aducanumab have been tested in large-scale randomized controlled trials. However, their use remains debated because of limited therapeutic benefit and the risk of adverse effects such as amyloid-related imaging abnormalities (ARIA), including edema (ARIA-E) and hemorrhage (ARIA-H) (Tolar et al., 2023).

Several investigational drugs have entered clinical development in recent years, showing promising outcomes. For instance, CT1812, a brain-penetrant molecule identified by van Dyck et al., disrupts the interaction between Aβ oligomers and neuronal receptors, thereby preventing synaptic injury (van Dyck et al., 2019). Masupirdine, a potent and selective 5-HT6 receptor antagonist, has demonstrated pro-cognitive activity in preclinical animal studies (Nirogi et al., 2023) and has shown beneficial effects on cognitive performance in clinical settings as well (Nirogi et al., 2022). Another notable example is sodium oligomannate (GV-971), which inhibits Aβ aggregation, facilitates the breakdown of existing aggregates, and reduces Aβ-induced neurotoxicity. Clinical evidence indicates that GV-971 can improve, and sometimes even reverse, cognitive deficits in AD patients (Syed, 2020; Klose et al., 2022).

3.2 Diagnostic Approaches for Alzheimer's Disease-Associated Cognitive Disabilities

Early and accurate diagnosis of AD is essential for timely intervention, management, and participation in clinical trials. Diagnostic evaluation generally involves clinical assessment, neuropsychological testing, neuroimaging, and biomarker analysis. Clinical assessment and cognitive testing are the primary foundations for diagnosis in research and clinical practice.

3.2.1 Clinical Assessment and Cognitive Tests

The diagnostic process begins with a comprehensive medical history and physical examination. Clinicians evaluate the onset, duration, and progression of cognitive and functional impairments and psychiatric or behavioral symptoms. Informant-based reports from caregivers are crucial, as patients may underreport memory difficulties or lack insight into their deficits. Structured instruments such as the Clinical Dementia Rating (CDR) and Global Deterioration Scale (GDS) help in staging disease severity (Scheltens et al., 2021).

Dementia is a growing global health concern, affecting over 55 million people worldwide, with numbers projected to reach 78 million by 2030 (Gauthier et al., 2021). Mild cognitive impairment (MCI), present in 12–18% of adults over 60, is a transitional stage between normal aging and dementia. Individuals with MCI have a higher risk of developing dementia, with annual progression rates of 8–18% (Gaugler et al., 2021; Petersen et al., 2018).

Current FDA-approved drugs for AD mainly provide temporary symptomatic relief. Recently approved agents, such as aducanumab and lecanemab, may slow disease progression, but their long-term impact remains uncertain (Esang & Gupta, 2021; van Dyck et al., 2022). Non-pharmacological strategies—including cognitive training, lifestyle modification, and psychosocial support—are being studied, though further evidence is needed (Arvanitakis et al., 2019). Among emerging therapies, transcranial magnetic stimulation has shown promise in improving cognitive functions in MCI and AD. Clinical studies and meta-analyses suggest potential benefits, but findings are limited by methodological variability, including differences in stimulation parameters and outcome measures (Birba et al., 2017;

Cheng et al., 2018; Dong et al., 2018). More standardized research is required to confirm its clinical utility.

3.2.2 Neuroimaging and Biomarkers

AD, the leading cause of dementia, is a progressive neurodegenerative disorder marked by Aβ plaques and intracellular NFTs composed of hyperphosphorylated tau, both regarded as pathological hallmarks of the condition (Sun et al., 2025). Impaired production or clearance of Aβ peptides results in amyloid deposition, which subsequently drives tau hyperphosphorylation, microglial activation, neurodegeneration, and cognitive decline (Hardy & Higgins, 1992).

To improve diagnostic precision, the National Institute on Aging has recommended a biologically based framework using the amyloid, tau, and neurodegeneration classification system for biomarker assessment (Jack et al., 2016). Measurement of these biomarkers in vivo is increasingly applied in clinical practice (Hansson et al., 2023). Recent updates suggest expanding biomarker categories to capture additional mechanisms such as neuroinflammation and vascular injury. However, their interpretation is still difficult to achieve in heterogeneous populations, particularly in individuals with non-Down syndrome intellectual disability. Biomarkers are measurable biological indicators to assess normal physiological states, pathological processes, or responses to therapeutic interventions. Neuroimaging biomarkers specifically refer to quantifiable parameters derived from imaging modalities such as magnetic resonance imaging (MRI), functional MRI (fMRI), positron emission tomography (PET), and diffusion tensor imaging (DTI) (Young et al., 2020). These biomarkers are significant in the early diagnosis of dementia, as they help identify characteristic structural, functional, and molecular changes (Hirsch et al., 2015; Tartaglia et al., 2011; Varghese et al., 2013). Volumetric MRI enables the detection of early structural alterations in the transentorhinal cortex and hippocampus, which are key in distinguishing AD from normal aging. Similarly, fMRI and DTI can reveal functional connectivity and white matter integrity disruptions, notably in regions such as the bilateral precuneus and corpus callosum. Molecular imaging further provides insights into altered metabolite concentrations in the posterior cingulate cortex, serving as potential diagnostic indicators for AD and MCI (Talwar et al., 2021).

Beyond AD, neuroimaging approaches, including MRI, single-photon emission computed tomography, PET, dopamine transporter scans, iodine-123 meta-iodobenzylguanidine myocardial scintigraphy, and amyloid and tau PET imaging, are critical in differentiating various neurodegenerative disorders (Table 3.1). These techniques can detect disease-specific patterns such as cortical atrophy, regional hypometabolism, and pathological protein accumulation, which are vital for diagnosing conditions like AD and dementia with Lewy bodies (Shimizu et al., 2018). More recent evidence also suggests that engagement in healthy daily practices is linked with favorable changes in neuroimaging markers, indicating potential protective effects on brain aging and reduced neurodegenerative progression (Townsend et al., 2022).

Table 3.1 Neuroimaging and biomarkers in AD diagnosis

Modality/Biomarker	Target/Function	Findings in AD	Clinical relevance	References
MRI	Brain structure	Hippocampal atrophy, cortical thinning	Early detection; excludes other causes of dementia	Jack et al., 2016
CT	Brain structure	Generalized cortical atrophy	Alternative when MRI not available	Scheltens et al., 1992
Fluorodeoxyglucose (FDG)-PET	Glucose metabolism	Hypometabolism in temporoparietal regions	Detects functional decline before structural atrophy	Mosconi, 2005
Amyloid PET	Amyloid plaques	Positive uptake in cortical regions	Supports early diagnosis; widely used in research	Klunk et al., 2004
CSF Aβ42	Amyloid pathology	Reduced concentrations	Differentiates AD from other dementias	Blennow & Zetterberg, 2018
CSF tau/p-tau	Tau pathology	Elevated total tau and phosphorylated tau	Correlates with disease progression and severity	de la Chapelle & Hampel, 2010

3.3 Therapeutic Management of Alzheimer's Disease-Associated Cognitive Disabilities

3.3.1 Pharmacological Intervention

Cholinesterase inhibitors: ChEIs—chiefly donepezil, rivastigmine, and galantamine—are first-line symptomatic treatments for mild-to-moderate AD (Birks et al., 2006). They act by inhibiting acetylcholinesterase, thereby augmenting cholinergic neurotransmission in cortical and hippocampal circuits implicated in memory and attention (Battle et al., 2021). Randomized trials and meta-analyses show modest but statistically significant improvements (or slower decline) in cognition, global clinical status, and activities of daily living for many patients on ChEIs compared with placebo; effect sizes are typically small to moderate and vary by drug and dose. Common adverse effects are cholinergic (nausea, vomiting, diarrhea, bradycardia) and may limit tolerability; gradual dose titration and monitoring for bradyarrhythmia and gastrointestinal side effects are recommended. Long-term observational and pooled analyses suggest that ChEIs may delay clinical decline in some patients when continued, but individual response is heterogeneous, and benefits should be reviewed periodically (Dou et al., 2018).

Clinical practice points (ChEIs): Start in mild-to-moderate AD when goals are symptom stabilization; titrate per product label; monitor tolerability (gastrointestinal symptoms, weight loss, syncope); reassess cognitive/functional benefit every

3–6 months; and consider deprescribing if no benefit or intolerable adverse effects (Bittner et al., 2023).

N-methyl-D-aspartate (NMDA) receptor antagonists: Memantine (an NMDA receptor antagonist) is indicated for moderate-to-severe AD and may be used alone or combined with a ChEI. Clinical trials indicate modest benefits for cognition, global status, and activities of daily living in moderate-to-severe disease; benefits for mild disease are not well supported. Memantine is generally well tolerated; common side effects include dizziness, headache, and constipation. Evidence supports combined memantine plus donepezil in selected patients with more advanced disease to stabilize function and behavior (Li et al., 2019; Xu et al., 2021). Clinical practice points (memantine): Consider moderate-to-severe AD, particularly where behavioral symptoms or functional decline are prominent; monitor for common adverse effects and drug interactions (Glinz et al., 2019).

Table 3.2 summarizes the class of pharmacological agents employed for ADCD management.

Recent years have seen advances in therapies targeting core AD pathologies (amyloid-β, tau). Monoclonal antibodies directed at amyloid (e.g., lecanemab and donanemab) have demonstrated reductions in brain amyloid burden and slight but measurable slowing of cognitive and functional decline in early symptomatic AD (prodromal/mild stages) in phase III trials (Yaghmaei et al., 2024). However, these agents carry risks, notably ARIA manifesting as edema or microhemorrhages; careful patient selection, MRI monitoring, and risk–benefit discussion are essential. Regulatory approvals and availability vary by jurisdiction and are typically limited to early AD in patients meeting biomarker criteria and under monitoring protocols (Jin & Noble, 2024). Ongoing trials clarify long-term efficacy, safety, and optimal patient selection. Clinical practice points (disease-modifying therapies): Restrict use to patients who match trial eligibility (early symptomatic AD with biomarker evidence where required); counsel patients/caregivers about modest clinical benefit, risks (ARIA), monitoring requirements (serial MRI), and costs; and refer to specialist centers when available (Rabinovici et al., 2024).

Table 3.2 Pharmacological management of ADCD

Drug/Class	Mechanism	Examples	Clinical effects	Adverse effects
ChEIs	Increase acetylcholine availability	Donepezil, rivastigmine, galantamine	Improve cognition, daily function	Nausea, vomiting, bradycardia
NMDA receptor antagonist	Reduce excitotoxicity	Memantine	Slows decline, improves behavior	Dizziness, confusion
Anti-amyloid antibodies	Clear amyloid plaques	Aducanumab, lecanemab	Potential disease modification	ARIA
Tau-targeting therapies	Reduce tau aggregation	Experimental	Under investigation	Not established

3.3.2 *Non-Pharmacological Interventions*

Cognitive Rehabilitation Cognitive rehabilitation (CR) refers to individualized, goal-directed interventions that teach compensatory strategies, restore lost skills, and train functional tasks relevant to everyday life. High-quality· systematic reviews and randomized trials indicate CR can improve performance in targeted daily activities and may enhance quality of life for people with mild-to-moderate dementia (Kudlicka et al., 2023). Evidence strength is moderate: Benefits are task-specific (i.e., most significant for activities directly trained), and sustained gains require ongoing practice or booster sessions. CR is most effective when integrated with caregiver training and environmental supports (Ren et al., 2024).

Practical recommendations: Set functional, person-centered goals with patient and caregiver; focus training on specific everyday tasks (e.g., medication management, route finding); combine strategy training with external aids (calendars, labeling); and measure success by real-world performance, not only neuropsychological tests (Xiao et al., 2024).

Physical Exercise Regular physical activity, predominantly aerobic and combined aerobic-resistance programs, has been associated with modest improvements or slower decline in cognition, mobility, and activities of daily living in people with AD and other dementias (Kouloutbani et al., 2022). Exercise may also reduce neuropsychiatric symptoms (e.g., apathy, depression) and improve physical function and independence. Different exercise modalities (aerobic, resistance, balance) show benefit; optimal dose and long-term adherence remain active areas of study. Exercise interventions are generally safe when individually tailored and supervised as needed (de Azevedo et al., 2021).

Practical recommendations: Aim for multimodal programs (aerobic + resistance + balance), 2–5 sessions per week as tolerated; tailor intensity to fitness/comorbidities; incorporate enjoyable activities to support adherence; and monitor cardiovascular safety in patients with comorbid disease.

Environmental Modifications Environmental design and targeted environmental interventions (cueing, orientation aids, improved lighting, noise reduction, simplified layouts, and signage) reduce confusion, agitation, and functional disability and support independence and safety for people with AD. Evidence from systematic reviews supports the role of dementia-friendly design principles in residential and home settings to reduce behavioral symptoms and improve orientation and performance of daily tasks. Simple, low-cost changes (consistent routines, prominent clocks/calendars, contrast cues for toileting, transparent floor surfaces) are practical and often effective (Quirke, 2023).

Practical recommendations: Assess the person's home (or care environment) to identify hazards and barriers; apply evidence-based design principles (clear sightlines, reduced clutter, good lighting, salient cues for key locations); provide staff/caregiver education on environmental strategies; and periodically reassess as the

Table 3.3 Non-pharmacological interventions in ADCD

Intervention	Method/Approach	Expected benefits	Evidence level
Cognitive rehabilitation	Memory training, orientation therapy	Improved cognition, daily function	Moderate
Physical exercise	Aerobic, resistance training	Improved blood flow, neuroplasticity	Strong
Environmental modifications	Safety adaptations, structured routines	Reduced agitation, better independence	Moderate
Psychosocial interventions	Social engagement, music/art therapy	Reduced depression, enhanced well-being	Growing evidence

disease progresses (Dementia Australia/practice guidance on environmental and non-pharmacological interventions).

Table 3.3 represents the non-pharmacological strategies for ADCD management.

Cholinesterase inhibitors (donepezil, rivastigmine, galantamine) and memantine remain core symptomatic pharmacotherapies with modest efficacy; monitor for adverse effects and individual response. Amyloid-targeting monoclonal antibodies (e.g., lecanemab, donanemab) represent the first wave of disease-modifying treatments for early AD but require careful selection, MRI monitoring for ARIA, and informed consent about modest benefits and risks (Vu et al., 2024). Non-pharmacological interventions—CR, structured physical exercise, and environmental modifications—are effective, low-risk strategies that improve functional outcomes and should be integrated into routine care (Guzzon et al., 2023). A multidisciplinary, person-centered approach that combines pharmacological and non-pharmacological measures produces the best overall outcomes for people living with AD-associated cognitive disabilities (ADCD).

3.4 Role of Caregivers in Alzheimer's Disease-Associated Cognitive Disabilities Management

Caregivers play a central role in the management of ADCD. Because AD is progressive, patients increasingly depend on informal caregivers (usually family members) or professional care staff to support daily living, manage behavioral disturbances, and coordinate healthcare. Effective caregiver engagement improves patient outcomes, reduces institutionalization, and helps preserve quality of life for both patient and family (Adelman et al., 2014; Gitlin et al., 2003).

3.4.1 Counseling and Support Systems

Counseling and structured support programs educate caregivers about the disease, coping strategies, stress management, and resource access. Evidence shows that caregiver burden—encompassing emotional distress, depression, financial strain, and social isolation—is one of the strongest predictors of early institutionalization of patients with AD (Brodaty & Donkin, 2009; Etters et al., 2008). Psychosocial interventions such as support groups, respite care, and counseling have been shown to reduce caregiver stress, delay nursing home placement, and improve mental health. Multicomponent interventions combining education, problem-solving therapy, and peer support are particularly effective (Schulz & Martire, 2004). Digital platforms and telehealth programs are emerging as accessible psychoeducation and psychological support tools, especially for rural or resource-limited populations (Mittelman et al., 2006; Boots et al., 2014).

3.4.2 Training and Family Involvement

Caregiver training equips family members with practical skills to manage cognitive, functional, and behavioral challenges in AD. Structured training programs can improve communication, reduce neuropsychiatric symptoms, and enhance adherence to care plans (Livingston et al., 2020). Family involvement in CR, exercise routines, and environmental modifications significantly improves adherence and functional outcomes (Moniz-Cook et al., 2008). Training caregivers to use behavioral management strategies (e.g., distraction, routine setting, validation therapy) reduces agitation and aggression in patients and lessens caregiver burnout (Clare & Woods, 2004). Furthermore, engaging families in shared decision-making regarding pharmacological treatments, lifestyle interventions, and advanced care planning fosters collaboration, improves satisfaction with care, and aligns treatment goals with patient values (Jennings et al., 2017).

3.5 Multidimensional Approaches for Improved Quality of Life

ADCD affects memory, executive function, emotional well-being, independence, and social participation. Because no single therapy meets all patient needs, a multidimensional approach is essential for improving quality of life. Pharmacological options such as ChEIs and NMDA receptor antagonists provide symptomatic relief, while nondrug therapies—including cognitive training, physical activity, music, and occupational therapy—help maintain function and reduce behavioral symptoms (Briks et al., 2006; Livingston et al., 2020). Psychosocial interventions like reminiscence therapy, group activities, and caregiver education support emotional well-being and social engagement (Olazarán et al., 2010). Environmental adaptations, such as structured routines and safe home designs, enhance autonomy and reduce caregiver strain. Effective care also relies on multidisciplinary collaboration among healthcare professionals, active caregiver participation, and community support systems, which sustain quality of life and lessen caregiver burden (Prince et al., 2016).

3.6 Summary and Conclusion

ADCD poses a complex challenge, as it affects not only memory and cognition but also daily functioning, emotional health, and social relationships. Management of ADCD requires a comprehensive approach that spans diagnosis, treatment, and long-term support. Early identification through clinical assessments, cognitive testing, neuroimaging, and biomarkers provides the foundation for timely intervention. Therapeutic management integrates pharmacological options, such as ChEIs and emerging agents, with non-pharmacological strategies including CR, structured exercise, and environmental modifications to preserve function and independence. Caregivers play a pivotal role, and their effectiveness is strengthened through counseling, support systems, and targeted training that encourage family involvement in care. Beyond individual measures, multidimensional approaches—combining medical, psychosocial, rehabilitative, and community resources—are essential to improving quality of life for patients while easing caregiver burden. Taken together, these strategies highlight the need for coordinated, patient-centered, and multidisciplinary care to address the complex challenges of ADCD.

References

Abbott, A. (2023). Are new Alzheimer's drugs really a breakthrough? *Nature, 616*, 20–22. https://doi.org/10.3390/molecules28227631

Adelman, R. D., Tmanova, L. L., Delgado, D., Dion, S., & Lachs, M. S. (2014). Caregiver burden: A clinical review. *JAMA, 311*(10), 1052–1060.

Arvanitakis, Z., Shah, R. C., & Bennett, D. A. (2019). Diagnosis and management of dementia: Review. *JAMA, 322*(16), 1589–1599. https://doi.org/10.1001/jama.2019.4782

AsdtjotAs. (2018). *World Alzheimer's disease report 2018*. Alzheimer's Disease International.

Battle, C. E., Abdul-Rahim, A. H., Shenkin, S. D., Hewitt, J., Quinn, T. J. (2021). Cholinesterase inhibitors for vascular dementia and other vascular cognitive impairments: Systematic review. *Journal of Neurology, Neurosurgery, and Psychiatry*. https://www.ncbi.nlm.nih.gov/pmc/articles/PMC8407366/

Birba, A., Ibáñez, A., Sedeño, L., Ferrari, J., García, A. M., & Zimerman, M. (2017). Non-invasive brain stimulation for the treatment of behavioral and cognitive symptoms in neurodegenerative diseases: A systematic review. *Frontiers in Neuroscience, 11*, 646. https://doi.org/10.3389/fnins.2017.00646

Birks, J. (2006). Cholinesterase inhibitors for Alzheimer's disease. *Cochrane Database Systematic Reviews, 1*, CD005593. DOI: 10.1002/14651858.CD005593.pub2. https://pubmed.ncbi.nlm.nih.gov/16437532/

Bittner, N., et al. (2023). A systematic review of AChEIs: Efficacy and adverse events. *Drugs & Aging*. https://doi.org/10.1007/s40266-023-01065-x

Blennow, K., & Zetterberg, H. (2018). Biomarkers for Alzheimer's disease: current status and prospects for the future. *Journal of Internal Medicine, 284*(6), 643–663. https://doi.org/10.1111/joim.12816

Boots, L. M., de Vugt, M. E., van Knippenberg, R. J., Kempen, G. I., & Verhey, F. R. (2014). A systematic review of Internet-based supportive interventions for caregivers of patients with dementia. *International Journal of Geriatric Psychiatry, 29*(4), 331–344.

Brodaty, H., & Donkin, M. (2009). Family caregivers of people with dementia. *Dialogues in Clinical Neuroscience, 11*(2), 217–228. https://doi.org/10.31887/DCNS.2009.11.2/hbrodaty

de la Chapelle, A., & Hampel, H. (2010). Clinical relevance of microsatellite instability in colorectal cancer. *Journal of Clinical Oncology: Official Journal of the American Society of Clinical Oncology, 28*(20), 3380–3387. https://doi.org/10.1200/JCO.2009.27.0652

Cheng, C. P. W., Chan, P., & Lee, T. M. C. (2018). Effects of repetitive transcranial magnetic stimulation on Chinese patients with Alzheimer disease: A randomized, double-blind, sham-controlled study. *Alzheimer Disease and Associated Disorders, 32*(3), 216–222. https://doi.org/10.1097/WAD.0000000000000253

Clare, L., & Woods, R. T. (2004). Cognitive training and cognitive rehabilitation for people with early-stage Alzheimer's disease: A review. *Neuropsychological Rehabilitation, 14*(4), 385–401. https://doi.org/10.1080/09602010443000074

de Azevedo, M. C. D., et al. (2021). Environmental interventions to support orientation and reduce behavioural symptoms in people with dementia: Systematic review. *Journal of Clinical Nursing*. https://doi.org/10.1590/1980-57642021dn15-040012

Dong, X., Yan, L., Huang, L., Guan, X., Dong, C., Tao, H., et al. (2018). Repetitive transcranial magnetic stimulation for the treatment of Alzheimer's disease: A systematic review and meta-analysis of randomized controlled trials. *PLoS One, 13*(10), e0205704. https://doi.org/10.1371/journal.pone.0205704

Dou, K. X., Tan, M. S., Tan, C. C., Cao, X. P., Hou, X. H., Guo, Q. H., Tan, L., Mok, V., & Yu, J. T. (2018). Comparative safety and effectiveness of cholinesterase inhibitors and memantine for Alzheimer's disease: A network meta-analysis of 41 randomized controlled trials. *Alzheimer's Research & Therapy, 10*(1), 126. https://doi.org/10.1186/s13195-018-0457-9

Esang, M., & Gupta, S. (2021). Aducanumab: An update on the new treatment of Alzheimer's disease. *Cureus, 13*(10), e18959. https://doi.org/10.7759/cureus.18959

Etters, L., Goodall, D., & Harrison, B. E. (2008). Caregiver burden among dementia patient caregivers: A review of the literature. *Journal of the American Academy of Nurse Practitioners, 20*(8), 423–428.

Gaugler, J. E., Roberts, L. A., & Gitlin, L. N. (2021). Dementia care and service utilization in the United States: Epidemiology and disparities. *Alzheimer's & Dementia, 17*(3), 456–467. https://doi.org/10.1002/alz.12205

Gauthier, S., Rosa-Neto, P., Morais J. A., Webster C. (2021). World Alzheimer Report 2021: Journey through the diagnosis of dementia. Alzheimer's Disease International. https://www.alzint.org/resource/world-alzheimer-report-2021/

Giacobini, E., Cuello, A. C., & Fisher, A. (2022). Reimagining cholinergic therapy for Alzheimer's disease. *Brain, 145*(7), 2250–2275. https://doi.org/10.1093/brain/awac096

Gitlin, L. N., Belle, S. H., Burgio, L. D., Czaja, S. J., Mahoney, D., Gallagher-Thompson, D., et al. (2003). Effect of multicomponent interventions on caregiver burden and depression: The REACH II randomized trial. *Gerontologist, 43*(4), 508–520.

Glinz, D., et al. (2019). Acetylcholinesterase inhibitors combined with memantine: Systematic review and meta-analysis. *Swiss Medical Weekly*. https://doi.org/10.4414/smw.2019.20093

Guzzon, A., Rebba, V., Paccagnella, O., Rigon, M., & Boniolo, G. (2023). The value of supportive care: A systematic review of cost-effectiveness of non-pharmacological interventions for dementia. *PLoS One*, *18*(5), e0285305. https://doi.org/10.1371/journal.pone.0285305

Hansson, O., Blennow, K., Zetterberg, H., & Dage, J. (2023). Blood biomarkers for Alzheimer's disease in clinical practice and trials. *Nature Aging*, *3*(5), 506–519. https://doi.org/10.1038/s43587-023-00403-3

Hardy, J. A., & Higgins, G. A. (1992). Alzheimer's disease: The amyloid cascade hypothesis. *Science*, *256*(5054), 184. https://doi.org/10.1126/science.1566067

Hirsch, G. V., Bauer, C. M., & Merabet, L. B. (2015). Using structural and functional brain imaging to uncover how the brain adapts to blindness. *Annals of Neuroscience and Psychology*, *2*, 5.

Jack, C. R., Jr, Bennett, D. A., Blennow, K., Carrillo, M. C., Feldman, H. H., Frisoni, G. B., Hampel, H., Jagust, W. J., Johnson, K. A., Knopman, D. S., & Petersen, R. C. (2016). A/T/N: An unbiased descriptive classification scheme for Alzheimer disease biomarkers. *Neurology*, *87*(5), 539–547. https://doi.org/10.1212/WNL.0000000000002923

Jennings, L. A., Palimaru, A., Corona, M. G., Cagigas, X. E., Ramirez, K. D., Zhao, T., et al. (2017). Informed family member involvement to improve the quality of dementia care in the health system. *Journal of the American Geriatrics Society*, *65*(7), 1520–1526. https://doi.org/10.1111/jgs.14299

Jin, M., & Noble, J. M. (2024). What's in it for me? Contextualizing the potential clinical impacts of lecanemab, donanemab, and other anti-β-amyloid monoclonal antibodies in early Alzheimer's disease. *Eneuro*, *11*(9). https://doi.org/10.1523/ENEURO.0088-24.2024

Kamble, S. M., Patil, K. R., & Upaganlawar, A. B. (2024). Etiology, pathogenesis of Alzheimer's disease and amyloid beta hypothesis. In Alzheimer's disease and advanced drug delivery strategies (pp. 1–11). Academic Press. https://doi.org/10.1016/B978-0-443-13205-6.00022-4

Klose, J., Lewe, J., & Pahnke, J. (2022). Sodium oligomannate (GV-971): A new multi-target drug candidate for Alzheimer's disease. *Neural Regeneration Research*, *17*(9), 1886–1888. https://doi.org/10.4103/1673-5374.335135

Klunk, W. E., Engler, H., Nordberg, A., Wang, Y., Blomqvist, G., Holt, D. P., Bergström, M., Savitcheva, I., Huang, G. F., Estrada, S., Ausén, B., Debnath, M. L., Barletta, J., Price, J. C., Sandell, J., Lopresti, B. J., Wall, A., Koivisto, P., Antoni, G., Mathis, C. A., … Långström, B. (2004). Imaging brain amyloid in Alzheimer's disease with Pittsburgh Compound-B. *Annals of Neurology*, *55*(3), 306–319. https://doi.org/10.1002/ana.20009

Kouloutbani, K., et al. (2022/2023). Effectiveness of exercise interventions on neuropsychiatric symptoms of dementia: Systematic review. *International Psychogeriatrics*.

Kudlicka, A., et al. (2023). Cognitive rehabilitation for people with mild to moderate dementia. *Cochrane Database of Systematic Reviews*, CD013388. https://doi.org/10.1002/14651858 CD013388.pub2.

Lashley, T., Schott, J. M., Weston, P., Murray, C. E., Wellington, H., Keshavan, A., et al. (2018). Molecular biomarkers of Alzheimer's disease: Progress and prospects. *Disease Models & Mechanisms*, *11*(5), dmm031781. https://doi.org/10.1242/dmm.031781

Li, D. D., et al. (2019). Meta-analysis: Memantine efficacy in Alzheimer's disease. *Frontiers in Neuroscience*, *13*, 472. https://doi.org/10.3389/fnins.2019.00472

Livingston, G., Huntley, J., Sommerlad, A., Ames, D., Ballard, C., Banerjee, S., et al. (2020). Dementia prevention, intervention, and care: 2020 report of the Lancet Commission. *Lancet*, *396*(10248), 413–446. https://doi.org/10.1016/S0140-6736(20)30367-6

Mangialasche, F., Solomon, A., Winblad, B., Mecocci, P., & Kivipelto, M. (2010). Alzheimer's disease: Clinical trials and drug development. *The Lancet Neurology*, *9*(7), 702–716. https://doi.org/10.1016/S1474-4422(10)70119-8

Mittelman, M. S., Haley, W. E., Clay, O. J., & Roth, D. L. (2006). Improving caregiver well-being delays nursing home placement of patients with Alzheimer disease. *Neurology*, *67*(9), 1592–1599. https://doi.org/10.1212/01.wnl.0000242727.81172.91

Moghekar, A., Li, S., Lu, Y., Li, M., Wang, M. C., Albert, M., et al. (2013). CSF biomarker changes precede symptom onset of mild cognitive impairment. *Neurology, 81*(2), 175–180. https://doi.org/10.1212/01.wnl.0000435558.98447.17

Moniz-Cook, E., Vernooij-Dassen, M., Woods, R., Verhey, F., Chattat, R., De Vugt, M., et al. (2008). A European consensus on outcome measures for psychosocial intervention research in dementia care. *Aging & Mental Health, 12*(1), 14–29. https://doi.org/10.1080/13607860801919850

Mosconi, L. (2005). Brain glucose metabolism in the early and specific diagnosis of Alzheimer's disease. FDG-PET studies in MCI and AD. *European Journal of Nuclear Medicine and Molecular Imaging, 32*(4), 486–510. https://doi.org/10.1007/s00259-005-1762-7

Nirogi, R., Shinde, A., Bogaraju, N., Bhyrapuneni, G., & Jayarajan, P. (2022). Clinical evaluation of masupirdine, a selective 5-HT6 receptor antagonist, in Alzheimer's disease. *Journal of Alzheimer's Disease, 87*(1), 123–134. https://doi.org/10.3233/JAD-220228

Nirogi, R., Shinde, A., Gaggin, P., Bhyrapuneni, G., & Jayarajan, P. (2023). Masupirdine: A selective 5-HT6 receptor antagonist with pro-cognitive activity in preclinical models. *Neuropharmacology, 229*, 109509. https://doi.org/10.1016/j.neuropharm.2023.109509

Olazarán, J., Reisberg, B., Clare, L., Cruz, I., Peña-Casanova, J., Del Ser, T., et al. (2010). Non-pharmacological therapies in Alzheimer's disease: A systematic review. *Dementia and Geriatric Cognitive Disorders, 30*(2), 161–178.

Petersen, R. C., Lopez, O., Armstrong, M. J., Getchius, T. S. D., Ganguli, M., Gloss, D., et al. (2018). Practice guideline update summary: Mild cognitive impairment. *Neurology, 90*(3), 126–135. https://doi.org/10.1212/WNL.0000000000004826

Prince, M., Comas-Herrera, A., Knapp, M., Guerchet, M., & Karagiannidou, M. (2016). *World Alzheimer Report 2016: Improving healthcare for people living with dementia.* Alzheimer's Disease International.

Quirke, M. (2023). Environmental design for people living with dementia. *Design & Health/MDPI.* https://doi.org/10.3390/encyclopedia3030076

Rabinovici, G. D., et al. (2024–2025). Appropriate use recommendations for donanemab and related agents (recent review). *European Journal of Neurology/The Lancet Neurology*, See clinical guidance articles for monitoring (MRI, ARIA management). https://doi.org/10.1016/j.tjpad.2025.100150

Ren, S., et al. (2024). The effect of cognitive rehabilitation on daily functioning in Alzheimer's disease: Systematic review. *Frontiers in Neurology.* https://doi.org/10.3389/fneur.2024.1371298

Scheltens, P., Leys, D., Barkhof, F., Huglo, D., Weinstein, H. C., Vermersch, P., Kuiper, M., Steinling, M., Wolters, E. C., & Valk, J. (1992). Atrophy of medial temporal lobes on MRI in probable Alzheimer's disease and normal ageing: diagnostic value and neuropsychological correlates. *Journal of Neurology, Neurosurgery, and Psychiatry, 55*(10), 967–972. https://doi.org/10.1136/jnnp.55.10.967

Scheltens, P., De Strooper, B., Kivipelto, M., Holstege, H., Chételat, G., Teunissen, C. E., … Van der Flier, W. M. (2021). Alzheimer's disease. *The Lancet, 397*(10284), 1577–1590. https://doi.org/10.1016/S0140-6736(20)32205-4

Schulz, R., & Martire, L. M. (2004). Family caregiving of persons with dementia: Prevalence, health effects, and support strategies. *American Journal of Geriatric Psychiatry, 12*(3), 240–249. https://doi.org/10.1097/00019442-200405000-00002

Shah-Abadi, A., Moghadasi, A., Nasiri, E., Mahmoudi, J., & Akbarzadeh, A. (2023). Amyloid beta-targeted nanomedicines for Alzheimer's disease therapy: Recent advances and future prospects. *International Journal of Pharmaceutics, 633*, 122603.

Shimizu, S., Hirose, D., Hatanaka, H., Takenoshita, N., Kaneko, Y., Ogawa, Y., Sakurai, H., & Hanyu, H. (2018). Role of neuroimaging as a biomarker for neurodegenerative diseases. *Frontiers in Neurology, 9*, 265. https://doi.org/10.3389/fneur.2018.00265

Sun, X., Zhu, J., Li, R., Peng, Y., & Gong, L. (2025). The global research of magnetic resonance imaging in Alzheimer's disease: A bibliometric analysis from 2004 to 2023. *Frontiers in Neurology*, *15*, 1510522. https://doi.org/10.3389/fneur.2024.1510522

Syed, Y. Y. (2020). Sodium oligomannate: First approval. *Drugs*, *80*(5), 441–444. https://doi.org/10.1007/s40265-020-01277-0

Talwar, P., Kushwaha, S., Chaturvedi, M., & Mahajan, V. (2021). Systematic review of different neuroimaging correlates in mild cognitive impairment and Alzheimer's disease. *Clinical Neuroradiology*, *31*(4), 953–967. https://doi.org/10.1007/s00062-021-01057-7

Tartaglia, M. C., Rosen, H. J., & Miller, B. L. (2011). Neuroimaging in dementia. *Neurotherapeutics*, *8*(1), 82–92. https://doi.org/10.1007/s13311-010-0012-2

Tolar, M., Abushakra, S., & Sabbagh, M. (2023). The path forward in Alzheimer's disease therapeutics: Re-evaluating the amyloid cascade hypothesis. *Alzheimer's & Dementia*, *19*(1), 31–46. https://doi.org/10.1002/alz.12752

Townsend, R. F., Woodside, J. V., Prinelli, F., O'Neill, R. F., & McEvoy, C. T. (2022). Associations between dietary patterns and neuroimaging markers: A systematic review. *Frontiers in Nutrition*, *9*, 806006. https://doi.org/10.3389/fnut.2022.806006

van Dyck, C. H., Nygaard, H. B., Chen, K., Donohue, M. C., Raman, R., Rissman, R. A., Brewer, J. B., Koeppe, R. A., Chow, T. W., Rafii, M. S., & Gessert, D. (2019). Effect of AZD0530 on cerebral metabolic decline in Alzheimer disease: A randomized clinical trial. *JAMA Neurology*, *76*(10), 1219–1229.

van Dyck, C. H., Swanson, C. J., Aisen, P., Bateman, R. J., Chen, C., Gee, M., et al. (2022). Lecanemab in early Alzheimer's disease. *The New England Journal of Medicine*, *387*(14), 1389–1397. https://doi.org/10.1056/NEJMoa2212948

Varghese, T., Sheelakumari, R., James, J. S., & Mathuranath, P. S. (2013). A review of neuroimaging biomarkers of Alzheimer's disease. *Neurology Asia*, *18*(3), 239.

Vu, H. T., Nguyen, H. T., & Nguyen, A. T. (2024). Effectiveness of non-pharmacological interventions for dementia among the elderly: A randomized controlled trial. *Geriatrics*, *9*(2), 52. https://doi.org/10.3390/geriatrics9020052

Wang, J., Gu, B. J., Masters, C. L., & Wang, Y. J. (2022). A systemic view of Alzheimer disease—insights from amyloid-β metabolism beyond the brain. *Nature Reviews Neurology*, *18*(11), 611–624. https://doi.org/10.1038/nrneurol.2017.111

Xiao, Y., et al. (2024). Meta-analysis: Efficacy of physical exercise interventions in Alzheimer's disease. *Aging Clinical and Experimental Research*. https://doi.org/10.3389/fpubh.2024.1485807

Xu, H., et al. (2021). Long-term effects of cholinesterase inhibitors on cognitive decline and mortality in Alzheimer's disease. *Neurology*. https://doi.org/10.1212/WNL.0000000000011832

Yaghmaei, E., et al. (2024). Combined use of Donepezil and Memantine increases survival—Recent evidence and implications. *Communications Medicine (London)*, *4*, 27. https://doi.org/10.1038/s43856-024-00527-6

Young, P. N., Estarellas, M., Coomans, E., Srikrishna, M., Beaumont, H., Maass, A., Venkataraman, A. V., Lissaman, R., Jiménez, D., Betts, M. J., & McGlinchey, E. (2020). Imaging biomarkers in neurodegeneration: Current and future practices. *Alzheimer's Research & Therapy*, *12*(1), 49. https://doi.org/10.1186/s13195-020-00612-7

Chapter 4
Current Research on Alzheimer's Disease-Associated Cognitive Disabilities Management and Future Perspectives

Subhayan Das, Shayeri Chatterjee Ganguly, and Moumita Kundu

Abstract Although several diagnostic and therapeutic strategies are being employed for Alzheimer's disease-associated cognitive disability (ADCD) management, preventing or reversing cognitive dysfunctions in Alzheimer's disease (AD) patients is still a significant challenge. This chapter delves into current research to advance the therapeutic approaches for ADCDs management. A deeper understanding of AD mechanisms is vital to exploring novel treatment strategies for managing ADCDs in a better way. This chapter also aims to provide research-based advanced insights into molecular mechanisms and neurodegenerative pathways associated with AD progression. Moreover, it reviews the discovery of innovative treatments, including both pharmacological and non-pharmacological interventions that mitigate cognitive decline. Furthermore, this chapter highlights the emerging application of nanotechnology for better management of ADCDs. It also addresses the role of artificial intelligence and machine learning (AI–ML) for personalized therapy and predictive modeling for ADCDs. Additionally, this chapter emphasizes the future perspectives of ADCD management, aiming for new inspirational interdisciplinary research directions with a vision of enhanced patient outcomes.

S. Das
Department of Allied Health Sciences, Brainware University, Barasat, Kolkata, West Bengal, India

S. C. Ganguly
Department of Pharmaceutical Technology, Brainware University, Barasat, Kolkata, West Bengal, India

M. Kundu (✉)
Department of Pharmaceutical Technology, Brainware University, Barasat, Kolkata, West Bengal, India;
Center for Multidisciplinary Research and Innovations, Brainware University, Barasat, Kolkata, West Bengal, India
e-mail: moumitakundu1991@gmail.com

M. Kundu, S. Das (eds.), *Alzheimer's Disease-Associated Cognitive Disabilities*, SpringerBriefs in Modern Perspectives on Disability Research,
https://doi.org/10.1007/978-981-95-7121-5_4

Keywords Alzheimer's disease-associated cognitive disabilities (ADCDs) · Artificial intelligence and machine learning (AI–ML) · Neurodegenerative pathways · Novel therapeutics · Nanotherapeutics

4.1 Introduction

Current research focuses on a variety of Alzheimer's disease-associated cognitive disabilities (ADCDs) causal factors due to the condition's intricacy and the large number of potential risk factors involved. Targeting one or more agents associated with ADCD, several compounds in the developmental stage are intended to impede the disease's progression (Carr et al., 1997; Shanmugam et al., 2022). These possible targets include immunological responses, neuroinflammation, metabolic changes, beta-amyloid and tau protein buildup, and many other factors (Galle et al., 2020). Numerous insights into the biochemical pathways behind AD have been produced by recent scientific and clinical research (Zhou et al., 2024). Even with improvements in our knowledge of the molecular causes of AD, attempts to create potent medications or non-pharmacological treatments to stop, prevent, or delay the illness's progression have mostly failed. Whether they are successful or not, extensive clinical trials and pharmacological research are crucial because they help find new medications or get rid of ones that don't work, which helps to steer the fight against AD. Future methods should be improved and modified based on the results of "failed" clinical studies (Atri et al., 2013; Schneider et al., 2014).

Since treating severe AD, which involves repairing damaged neurons and synapses, will probably remain difficult, at least in the short term, priority should be given to preventing or slowing neurodegeneration, especially in those who are at risk. Modern technology-based non-pharmacological methods, such as minimally invasive or noninvasive surgery, ought to be actively incorporated into AD prevention plans (Ganguly et al., 2024). Through processes like neurogenesis, maintaining a healthy lifestyle that includes exercise, nutrition, and sleep can be advantageous and may help postpone the onset of Alzheimer's symptoms. For many senior people, AD could be eradicated if a substantial delay of one to two decades or longer can be accomplished. In addition to risk-reduction strategies similar to those used in present therapy for many cancer and AIDS cases, it is expected that the upcoming treatments will involve a combination of medications or devices that target various objectives. There was a tendency for slower tau buildup in donanemab clinical trials (Carr et al., 1997; Galle et al., 2020; Olczak et al., 2021; Sadiq, 2023; Se Thoe et al., 2021). Therefore, in order to provide significant benefits for Alzheimer's patients, future research should concentrate on examining the connection between decreased Aβ plaques and tau levels. Future patient therapies should be more efficient and reasonably priced thanks to the FDA's expedited approval of aducanumab, donanemab, and lecanemab, which has given Alzheimer's medication development fresh hope (Sadiq, 2023).

4.2 Current Research in Diagnostic and Therapeutic Strategies

With developments aimed at early detection and successful therapy intervention, advancements in AD diagnostics have been developed. Cognitive assessment, neuroimaging, and the screening of specific biomarkers are currently the main methods used to diagnose AD (Se Thoe et al., 2021; Sharma et al., 2025). The combination of cerebrospinal fluid (CSF) biomarkers and imaging was studied since it was possible to reliably identify the condition, ideally before symptoms appeared, using minimally invasive and inexpensive procedures (El Kadmiri et al., 2018). The development of methods to quantify Aβ in vivo—which is still the primary target of disease-modifying drug candidates in clinical trials—via amyloid positron emission tomography (PET) imaging and CSF biomarkers did, in fact, contribute to the molecular elucidation of the underlying mechanisms. By making it easier to track the pharmacodynamic effects of new drug candidates, this can eventually enhance the development of target therapy (Zetterberg & Bendlin, 2021). The optimal biomarker, which can be detected or measured in biological fluids such as blood, CSF, or peripheral tissue, should possess several essential characteristics, including high sensitivity and high specificity to enable differentiation. AD should be differentiated from other dementias with high reliability; it must possess a positive predictive value, be associated with AD pathology, facilitate early detection for the provision of diagnostic and prognostic information or optimization of treatment strategies, and be achievable through a noninvasive and cost-effective technique (Hansson, 2021). Table 4.1 illustrates the potential biomarkers for ADCD.

The treatment involves the use of acetylcholinesterase inhibitors (AChEIs) (donepezil, galantamine, and rivastigmine) and the N-methyl-D-aspartate (NMDA) receptor antagonist memantine, aiming to enhance cholinergic neurotransmission and activate NMDA receptors, respectively (Ismaili et al., 2017; Savelieff et al., 2019). Currently, the development of AD drugs encompasses disease-modifying strategies that concentrate on either immunotherapies or small molecules intended for oral administration (Cummings et al., 2018; Cummings et al., 2014; Ibrahim & Gabr, 2019; Weller & Budson, 2018). Tacrine, characterized by its neuroprotective and antioxidant properties, has been utilized to create hybrids with ferulic acid, caffeic acid, and Trolox. Trolox is a potent antioxidant that safeguards the liver from oxidative damage and exhibits neuroprotective effects by scavenging reactive oxygen species (ROS) and reducing Aβ-mediated neurotoxicity (Li et al., 2017). Recently, Namzaric, a combination of the previously authorized medications donepezil hydrochloride and memantine hydrochloride (extended-release) (Deardorff & Grossberg, 2016), was released for the treatment of AD. Monoclonal antibodies lecanemab and aducanumab, which can function as anti-amyloid antibodies to treat early-stage AD, have received FDA approval. According to Rahman et al., these antibodies are specifically designed for those with mild dementia or cognitive impairment brought on by AD, as well as those with proven high levels of Aβ in the early stages of AD (Rahman et al., 2020; Reardon, 2023; Wu et al., 2023).

Table 4.1 List of some potential biomarkers for ADCD

Sl. No.	Biomarkers	Characteristics
1	Aβ	Aggregation of Aβ in brain parenchyma leads to lower CSF levels and production of senile plaques (around 50% of normal Aβ42 concentration)—amyloid PET is the most commonly used biomarker in clinical trials, with three approved tracers for cognitive impairment evaluation—the ratio of peptide Aβ42 to soluble Aβ40 aggregation concentration is used to improve diagnosis precision, correlating approximately 100% with amyloid PET
2	Total and phosphorylated tau	CSF concentrations of total and phosphorylated tau are useful indicators of AD pathophysiology, reflecting neurodegenerative dementia pathophysiology. Increased phosphorylated tau levels indicate protein phosphorylation and neurofibrillary tangle (NFT) formation. Higher total tau levels indicate neuronal release, indicating increased neuronal activity. Studies indicate that plasma P-tau181 is strongly correlated with both CSF tau and amyloid PET levels
3	Dendritic protein neurogranin	Neurogranin, a biomarker for synaptic dysfunction and degeneration, is found in dendritic spines and correlates with cognitive decline in AD patients
4	PET tracers targeting synaptic vesicle protein 2A (SV2A)	Biomarker of synaptic loss: PET imaging facilitates the quantification of brain SV2A density. In AD patients, studies have shown a reduced density of SV2A in the hippocampus
5	Volumetric magnetic resonance imaging (MRI)	As a biomarker for AD neurodegeneration, it aids in diagnosing dementia patients. A weighted imaging method evaluates morphological changes in the brain and medial temporal volumes, including brain atrophy, gray matter atrophy, regional atrophy, and surface-based cortical thickness
6	Inflammation/astroglial activation	CSF levels of chitinase-3-like protein 1 (YKL-40), a glycoprotein found in astrocytes and microglia, and triggering receptor expressed on myeloid cells-2 (TREM2) are elevated in patients, showing a positive correlation with CSF levels of tau and Aβ. Glial fibrillary acidic protein (GFAP), a marker of astrocyte activation, is identified as a CSF biomarker
7	MicroRNA	Recent studies indicate that miRNA-125b concentrations are reduced in the serum of AD patients compared to noninflammatory controls, suggesting its potential as a biomarker for the disease

(continued)

Table 4.1 (continued)

Sl. No.	Biomarkers	Characteristics
8	Neurofilament light (NfL)	NfL serves as a biomarker for neuroaxonal damage and degeneration, measurable in both CSF and plasma/serum. It is a neuronal cytoplasmic protein found in myelinated axons. The levels of NfL in CSF and blood increase in relation to the extent of axonal damage. In familial AD, particularly among mutation carriers, reduced NfL levels can be observed years prior to the clinical onset of the disease
9	Fluorodeoxyglucose PET	A radiolabeled glucose molecule is utilized for AD diagnosis through the assessment of synaptic dysfunction. This technique evaluates the presence of lesions characterized by elevated glycolytic metabolism. Hypometabolism indicates a combination of synaptic and metabolic dysfunction, alongside neuronal cell loss. It is capable of detecting abnormalities in the early stages of the disease

4.3 Molecular Mechanisms and Neurodegenerative Pathways in AD

Recent research explored several new factors contributing to the disease progression of AD (Fig. 4.1). AD is commonly found to be associated with mitochondrial dysfunction. Mitochondrial dysfunction leads to ROS generation and causes defective mitophagy (Hein et al., 2025; Shi et al., 2025). Downregulation of Nrf2 has been shown to suppress mitochondrial biogenesis through reduced PPARγ–PGC1α signaling, causing mitochondrial loss, enhanced Aβ accumulation, synaptic damage, and cognitive deterioration in Amyloid Precursor Protein / Presenilin 1 (APP/PS1) models (Li et al., 2025; Luo et al., 2025). Transcriptomic studies and single-cell dataset analysis revealed metabolic reprogramming, stimulating fatty acid oxidation and glycolysis and leading to oxidative phosphorylation (Chu et al., 2025). Cell death pathways including apoptosis, ferroptosis, necroptosis, pyroptosis, anoikis, and NETosis cause synaptic dysfunction and neurovascular damage, contributing to AD pathology (Xu J. et al., 2025; Zeng & Jin, 2025). Hippo signaling is upregulated in ferroptosis mechanism, causing LATS1/2 stimulation, and this protein inhibits the genes involved in neuronal survival. LATS1/2 inhibitors have been found to improve the cognitive function in AD murine model (Evans et al., 2025). Another group found the correlation of ER stress response and AD. PERK (EIF2AK3), along with other transcription factors including FOSB and MEOX1, stimulates epithelial–mesenchymal transition (EMT), cell death pathways, and fibrosis as a part of endoplasmic reticulum (ER) stress responses. Inhibitors of these regulators have been identified as potential agents for AD therapy (Saptarshi et al., 2022; Shi et al., 2025). Moreover,

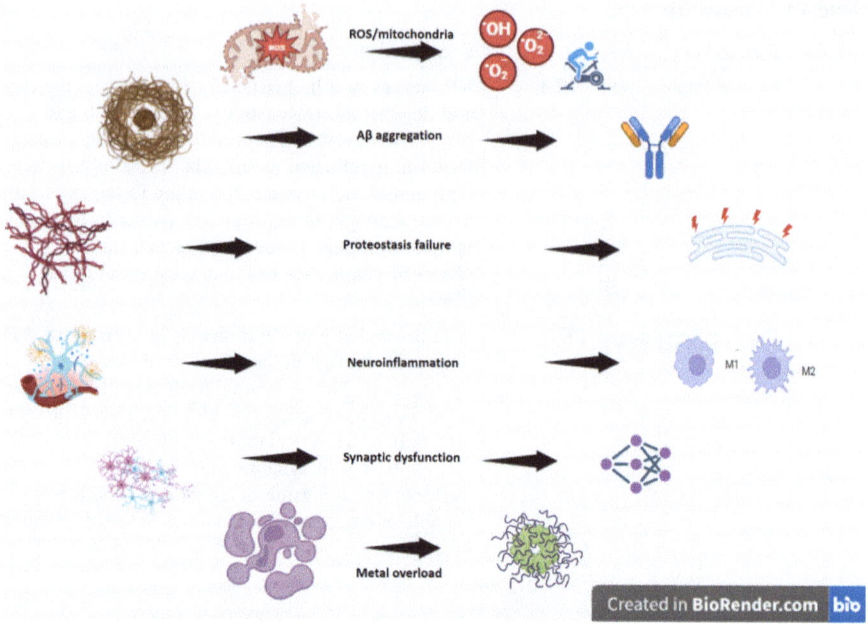

Fig. 4.1 Key mechanistic pathways and intervention strategies for ADCD. Mitochondrial dysfunction, amyloid-β aggregation, tau pathology, neuroinflammation, synaptic loss, and ferroptosis/apoptosis are core pathological pathways related to AD cognitive disorders. Intervention classes mapped to these pathways include antioxidants and exercise-based metabolic support; Aβ-directed antibodies; autophagy activators; anti-inflammatory modulators and M1→M2 microglial reprogramming agents; cognitive training and neurostimulation approaches; and metal chelators

hyperactivation of innate immune response due to Aβ accumulation stimulates TLR and NF-κB signalings and activates pro-inflammatory microglia. This leads to cytokine activation to cause apoptosis of neurons (Miao et al., 2023). Multiple studies confirm that chronic switching of microglia toward the pro-inflammatory M1 phenotype leads to release of TNF-α, IL-1β, IL-6, and upregulation of oxidative enzymes—a process now central to AD pathology (Gao et al., 2023). Abnormality in the retromer complex like VPS35, VPS29, and VPS26 also interferes with the homeostasis mechanism of microglia (Wu et al., 2024). Overstimulation of mTOR causes defective proteosis, resulting in Aβ accumulation and tau aggregation. Additionally, autophagic flux improves cognitive activity in AD (Davoody et al., 2024). Diet-derived bioactive peptides have also emerged as contributors to neuroprotective pathways. Hempseed protein-derived peptides demonstrated the ability to cross intestinal and blood–brain barriers and suppress hippocampal inflammatory markers such as GFAP, IBA1, TREM2, iNOS, COX-2, and IL-6 while enhancing IL-10 expression, positioning them as promising dietary immunomodulators in AD management (Hein et al., 2025; Torrecillas-Lopez et al., 2025).

4.4 Innovative Pharmacological Interventions

4.4.1 Disease-Modifying Agents

Researchers have reported novel AChE inhibitors targeting dual sites of the enzyme—the catalytic active site (CAS) and peripheral anionic site (PAS). These AChE inhibitors block acetylcholine degradation and Aβ accumulation to provide symptomatic relief and suppress disease progression (Gajendra et al., 2024). Monoclonal antibodies are also employed as anti-aggregation therapy in AD treatment. Lecanemab and donanemab slow down the cognitive dysfunction by lowering amyloid expression in the early stage. However, monoclonal antibodies possess several limitations such as high-cost, amyloid-related imaging abnormalities (Wang et al., 2025). Hybrid molecules using multi-target-directed ligands are also being employed for ADCD. Cholinesterase inhibitors (ChEIs) are combined with various antioxidants, anti-amyloids, and metal-chelating agents to target multiple pathways of AD (Agis-Torres et al., 2014). Some small molecules that stabilize VPS35 are also being studied as these are able to reduce Aβ deposition, leading to normalization of AMPA receptor signaling, calcium homeostasis restoration (Ramonet et al., 2025). Researchers also reported several natural products such as ajugol and isoacteoside for ADCD management. Ajugol, *Rehmannia glutinosa* derivative, inhibits TLR/NF-κB/NLRP3 pathway and caspase-3-mediated apoptosis. On the other hand, isoacteoside stimulates autophagy through LC3/P62/p-mTOR, resulting in cognitive function improvement (Han et al., 2025). Numerous flavonoids—including myricetin, scutellarin, hesperidin, rutin, and luteolin—attenuate glial activation, inhibit MAPK and NF-κB signaling, and reduce inflammatory cytokines (Al-Khayri et al., 2022). These natural product findings reinforce the emerging shift toward neuroinflammation-targeted AD therapies. Advanced strategies including lipidization, intranasal delivery, and light-activated inhibitors are also employed to improve blood–brain barrier penetration and target-specific drug action (Liu et al., 2025).

4.4.2 Neuroprotective Approaches

Neuroprotective approaches aim to improve neuronal viability and target the downstream pathology. Mitochondrial antioxidants reduce ROS-mediated apoptosis by maintaining the energy system. Antioxidants targeting mitochondrial health reduce ROS accumulation and restore bioenergetic stability, which is especially relevant given Nrf2-regulated mitochondrial biogenesis failures observed in AD (Li et al., 2025). Inflammation-modulatory agents can reprogram microglia by promoting pro-inflammatory M1 to neurotrophic M2 phenotype conversion (Guo et al., 2022). Transcription factor EB (TFEB) activators are also reported to boost Aβ/tau clearance by restoring lysosomal biogenesis (Lin et al., 2024). Moreover,

BCL-2/caspase cascades are targeted to suppress apoptosis of neurons. Several compounds target Hippo pathway to block ferroptosis, supporting neuronal resilience (Galvan et al., 2000). Nowadays, AI-assisted drug research, induced pluripotent stem cell (iPSC)-derived neuronal models, and advanced nanotechnology are being applied to accelerate neuroprotective drug discovery (Hein et al., 2025; Sheikh et al., 2025).

4.5 Non-Pharmacological Interventions Under Research

4.5.1 Cognitive Enhancement Therapies

Non-pharmacological therapies support the pharmacological treatment through cognitive function improvement. Several AD patients have benefitted from cognitive training and computerized memory exercises. These approaches can also slow down the memory decline process (Cui et al., 2018). Preclinical behavioral paradigms such as Morris water maze and open-field tests result in enhanced outcomes when combined with pharmacological therapy. Moreover, noninvasive brain stimulation techniques have been employed as adjunct therapy to enhance neuroplasticity. Many countries include structured cognitive rehabilitation in the recommended guidelines for early AD therapy (Fonte et al., 2001).

4.5.2 Lifestyle and Behavioral Interventions

Regular exercise improves mitochondrial health, lowers amyloid burden, and boosts synaptic plasticity (Fonte et al., 2001). A systematic review concluded that aerobic activity positively influences memory, executive function, metabolic health, and neuroinflammation in patients with neurocognitive disorders (Soldevila-Matías et al., 2024). Dietary modifications like antioxidant-rich, polyphenol-based, mediterranean diets, etc. aid in ADCD management by oxidative stress reduction. Some compounds like catalpol and verbascoside have shown anti-inflammatory and neurotrophic effects (Nani et al., 2021). Proper sleep and stress reduction help in mitigating systemic inflammation while promoting proteostasis (Fonte et al., 2001). Researchers revealed that gut microbiota modulation via highlighting prebiotics, probiotics, and dietary modulation leads to neuroprotection (Soldevila-Matías et al., 2024). Some countries like New Zealand, Australia, and Taiwan have taken dementia-friendly initiatives and have established the role of social inclusion in cognition enhancement (Nani et al., 2021).

A summary of current and emerging interventions for ADCD management is presented in Table 4.2.

Table 4.2 Current and emerging interventions in ADCD management

Intervention type	Mechanism of action	Preclinical/clinical status
Dual-site AChE inhibitors	Block catalytic hydrolysis at CAS + prevent AChE-mediated Aβ aggregation at PAS	Preclinical; rational drug design; limited clinical translation
Anti-aggregation therapies (antibodies, peptides, polyphenols)	Bind Aβ oligomers/fibrils, disrupt aggregation, promote clearance	Multiple antibodies in phase II/III trials (e.g., lecanemab, donanemab); small molecules mostly preclinical
Multi-target-directed ligands (MTDLs)	Hybrid compounds combining ChEI with antioxidant, anti-amyloid, and metal-chelating functions	Preclinical; some hybrids in early clinical evaluation
Natural products (ajugol, isoacteoside)	Ajugol: inhibits NF-κB/NLRP3 and BCL-2/BAX/caspase-3 apoptosis pathways; isoacteoside: enhances autophagy via LC3/P62/mTOR	Demonstrated efficacy in APP/PS1 and D-gal/$AlCl_3$ mouse models; validated in vitro
Metal chelators & antioxidants	Bind Cu^{2+}/Fe^{2+}, lower ROS, and reduce metal-induced Aβ aggregation	Preclinical; clioquinol derivatives reached clinical trials
Autophagy/proteostasis enhancers	Activate TFEB, promote lysosomal biogenesis, increase Aβ/tau clearance	Preclinical; isoacteoside as proof-of-concept
Neuroinflammation modulators	Suppress TLR/NF-κB/NLRP3 signaling; reprogram microglia from M1 (pro-inflammatory) to M2 (neurotrophic)	Ajugol validated in vivo and in vitro; several inhibitors under development
Cognitive training & brain stimulation (transcranial magnetic stimulation (TMS), transcranial direct current stimulation (tDCS))	Enhance neuroplasticity and compensatory circuits	Pilot human trials; preclinical animal behavioral studies
Lifestyle (exercise, diet, microbiome modulation)	Improve mitochondrial health, reduce amyloid burden, modulate gut–brain axis	Strong preclinical and epidemiological evidence; ongoing lifestyle intervention trials

4.6 Emerging Role of Nanotechnology in ADCD Management

Recent advancements in nanotechnology and nanomedicine have been achieved to enhance the permeability of some substances across the blood–brain barrier and to improve their stability and bioavailability. Curcumin-loaded nanoparticles are currently being evaluated to address the issues of low stability, bioavailability, and rapid oxidation of coumarin. Moreover, nanoparticles have demonstrated efficacy as a viable alternative to mitigate certain documented adverse effects of chelating treatment, such as hepatotoxicity, limited ability to traverse the blood–brain barrier, and neurotoxicity (Gupta et al., 2019; Liu et al., 2006). Ceria nanoparticles, alongside nanoparticles infused with Fe and Cu chelating agents and their derivatives, have been employed to safeguard mitochondria against ROS-induced oxidative stress. Likewise, gold- and selenium-containing nanoparticles have been investigated for their inhibitory effects on Aβ aggregation (Celardo et al., 2011; Kwon et al., 2016). Conversely, nanoparticles have been utilized in the diagnosis of AD. Ultra-small gadolinium nanoparticles demonstrated the capacity to bind to Aβ peptide. Furthermore, when modified with polyisobutylene (PIB) or a nanobody (B10AP) that interacts with Aβ plaques. Diagnosis of Alzheimer's Disease (Gupta et al., 2019). Gold, silver, and magnetic nanoparticles are being employed in AD diagnosis through the development of biosensors, which integrate a biochemical binding element with a signal conversion mechanism for the assessment of several AD biomarkers. Gold nanoparticles have exceptional chemical stability and conductivity, appropriate catalytic activity, extensive surface area, biological compatibility, and ease of functionality, enhancing the immobilization of biological recognition elements such as DNA, antibodies, and enzymes. These characteristics render them extensively investigated in the development of biosensors to authenticate AD biomarkers for detection, mostly owing to their capacity to bind to Aβ peptide (Bilal et al., 2020; Monteiro et al., 2023). Lipid nanoparticles (LNPs) infused with pomegranate extract were examined in a rat model of AD produced by aluminum chloride. It reduced NFTs and beta-amyloid (Aβ) accumulation (Almuhayawi et al., 2020). Erythropoietin (EPO)-encapsulated solid lipid nanoparticles (SLNs) reduce oxidative stress and amyloid-beta (Aβ) accumulation while enhancing spatial memory (Dara et al., 2019). Nanostructured lipid carriers (NLCs) demonstrate a strong affinity for Aβ and facilitate its breakdown (Gobbi et al., 2010). NLCs exhibit enhanced stability and possess a substantial loading capacity (Plaza-Oliver et al., 2021). NLCs facilitate drug transfer from the nasal cavity to the brain and are utilized in the treatment of AD related to microglial activation (Costa et al., 2021). Researchers produced poly(lactic-co-glycolic acid) (PLGA)-coated curcumin nanoparticles, demonstrating antioxidative properties and disaggregating Aβ clumps in an AD animal model (Mathew et al., 2012). Huperzine A was incorporated into PLGA coupled with lactoferrin nanoparticles, demonstrating improved release kinetics and markedly reduced AD symptoms (Meng et al., 2018). Zinc- and sitagliptin-loaded nanoparticles demonstrate enhanced cognitive performance and decreased neuroinflammation as anti-Alzheimer's benefits (Mir Najib Ullah et al., 2023; Vilella et al., 2018).

4.7 Artificial Intelligence and Machine Learning (AI–ML) in ADCD

Artificial intelligence and machine learning has become an integral part of research and development in recent years. A lot of effort has been dedicated toward integrating AI–ML in diagnosis and therapeutics in different diseases. In AD, integration of AI–ML has been challenging. Researchers are harnessing the power of AI–ML to develop early diagnostic modalities, innovate predictive analysis of disease progression, and curate personalized treatment options (Panwar et al., 2024).

4.7.1 AI–ML to Develop Early Diagnostic Modalities

Artificial Intelligence can assist doctors and researchers to enhance the existing diagnostic modalities which can aid in enhancing the output of existing diagnostic modalities. AI enhancement can assist in image enhancement leading to early detection of AD through MRI and PET scans. Deep learning systems and ML algorithms can be used to analyze these data to detect early structural changes in the brain structure that is otherwise unnoticed (Akan et al., 2025; Basanta-Torres et al., 2025; Kale et al., 2024). These models can also be trained to detect and analyze the blood parameters to predict the emergence of AD at an early age. Already research has been progressing to detect Alzheimer's associated proteins like tau and amyloid β for early detection of the disease (Akan et al., 2025; Kale et al., 2024; Panwar et al., 2024). AI–ML can also be utilized to develop tools to identify cognitive and behavioral abnormalities like speech and language abnormalities (Bergamasco et al., 2025).

4.7.2 AI–ML in Predictive Analysis of Disease Progression

AI–ML models can predict the progression and outcomes of AD by analyzing multimodal data like imaging, biomarkers, and clinical parameters. ML algorithms can be utilized to predict a particular outcome of the disease or a particular management option based on these data. Addition of genetic or epigenetic data can also assist in further refinement of these algorithms to predict the disease progression more accurately. Recently different approaches have been developed to analyze multimodal data. The multi-block alternating direction method integrates multimodal longitudinal clinical data to predict cognitive scores and diagnoses by utilizing the underlying structure of clinical information (Tabarestani et al., 2020). Tensor multitask learning predicts AD progression by analyzing spatiotemporal changes in brain biomarkers with tensor decomposition and gradient boosting (Zhang et al., 2023). Dynamic prediction with transformers leverages long-term follow-up data to better model time-based dependencies and enhance personalized AD progression predictions (Xu W. et al., 2025). Gradient boosting and ensemble learning increase prediction accuracy and stability by combining results from several models (Tabarestani et al., 2020).

4.7.3 AI–ML in Personalized Therapy Development for Alzheimer's Disease

AI–ML algorithms can enhance the diagnosis of AD, enhance the biomarker discovery, monitor disease progression, assist in drug development and repurposing, develop a clinical support system, and lastly integrate multimodal data for personalized interventions. We have discussed some of these aspects in previous sections. Here, we will be focusing on drug development and personalized interventions. AI speeds up drug discovery for AD by finding targets and predicting compound efficacy. It enables faster development and repurposing of treatments tailored to patients. AI-powered decision support tools help healthcare professionals make better choices for their patients (Kale et al., 2024). ML systems can suggest tailored treatment plans and keep track of patient progress by examining patient data, ensuring that care strategies keep on improving (Manubolu & Peeriga, 2025). AI tools like facial expression recognition and retinal imaging are able to enable noninvasive, real-time disease monitoring, allowing prompt treatment adjustments (Goudarzi et al., 2025). AI can personalize therapy by predicting patient responses, optimizing medications, and adjusting treatments. It also supports tailored nondrug interventions like cognitive training and rehabilitation to boost cognitive function and slow disease progression (Gu et al., 2025).

4.8 Conclusions and Future Perspectives

Alzheimer's Disease-associated Cognitive Disability (ADCD) is an increasing public health issue due to the presence of irreversible neurodegeneration in key areas of brain. While research has clarified mechanisms like amyloid and tau buildup, mitochondrial dysfunction, neuroinflammation, and synaptic loss, effective treatments remain elusive. Current therapeutic approaches are mostly directed to symptomatic relief rather than complete cure. Newer medications like anti-amyloid monoclonal antibodies offer limited benefits due to safety concerns, high costs, and narrow use. Disease-modifying drugs, neuroprotective agents, and advanced delivery systems like nanotechnology to improve blood–brain barrier penetration have shown promise and can be integrated to enhance current therapeutic approaches. However, lifestyle modifications and cognitive interventions still play a crucial role in early functional recovery.

AI and ML are transforming diagnostics and personalized medicine in AD. Still there is a long way to go to reach a meaningful improvement in current management. Enhanced imaging, biomarkers, and digital tools can be developed to aid early detection and intervention. AI can enhance drug discovery and support better, individualized treatment choices based on multimodal data analysis from different sources. In the future, approaches to precision medicine that combine genomics, proteomics, imaging, and patient clinical data may play a vital role in developing tailored therapeutic approaches. Building healthcare systems that are supportive of dementia and making innovative therapies affordable and accessible are essential for achieving this goal.

In summary, ADCD is still incurable, but new biomedical research and therapies provide hope for slowing its progression and improving patients' lives.

References

Agis-Torres, A., Sollhuber, M., Fernandez, M., & Sanchez-Montero, J. M. (2014). Multi-target directed ligands and other therapeutic strategies in the search of a real solution for Alzheimer's disease. *Current Neuropharmacology*, *12*(1), 2–36. https://doi.org/10.2174/1570159X113116660047

Akan, T., Alp, S., Bhuiyan, Md. S., Disbrow, E. A., Conrad, S. A., Vanchiere, J. A., Kevil, C. G., & Bhuiyan, M. A. N. (2025). Artificial intelligence-based framework for Alzheimer's disease diagnosis via video vision transformer. *Intelligent Medicine*. https://doi.org/10.1016/j.imed.2025.07.001

Al-Khayri, J. M., Sahana, G. R., Nagella, P., Joseph, B. V., Alessa, F. M., & Al-Mssallem, M. Q. (2022). Flavonoids as potential anti-inflammatory molecules: A review. *Molecules*, *27*(9), 2901. https://doi.org/10.3390/molecules27092901

Almuhayawi, M. S., Ramadan, W. S., Harakeh, S., Al Jaouni, S. K., Bharali, D. J., Mousa, S. A., & Almuhayawi, S. M. (2020). The potential role of pomegranate and its nano-formulations on cerebral neurons in aluminum chloride induced Alzheimer rat model. *Saudi Journal of Biological Sciences*, *27*(7), 1710–1716. https://doi.org/10.1016/j.sjbs.2020.04.045

Atri, A., Molinuevo, J. L., Lemming, O., Wirth, Y., Pulte, I., & Wilkinson, D. (2013). Memantine in patients with Alzheimer's disease receiving donepezil: New analyses of efficacy and safety for combination therapy. *Alzheimer's Research & Therapy*, *5*(1), 6. https://doi.org/10.1186/alzrt160

Basanta-Torres, S., Rivas-Fernández, M. Á., & Galdo-Alvarez, S. (2025). Artificial intelligence for Alzheimer's disease diagnosis through T1-weighted MRI: A systematic review. *Computers in Biology and Medicine*, *197*, 111028. https://doi.org/10.1016/j.compbiomed.2025.111028

Bergamasco, L., Coletta, A., Olmo, G., Cermelli, A., Rubino, E., & Rainero, I. (2025). AI-based facial emotion analysis for early and differential diagnosis of dementia. *Bioengineering*, *12*(10), 1082. https://doi.org/10.3390/bioengineering12101082

Bilal, M., Barani, M., Sabir, F., Rahdar, A., & Kyzas, G. Z. (2020). Nanomaterials for the treatment and diagnosis of Alzheimer's disease: An overview. *NanoImpact*, *20*, 100251. https://doi.org/10.1016/j.impact.2020.100251

Carr, D. B., Goate, A., Phil, D., & Morris, J. C. (1997). Current concepts in the pathogenesis of Alzheimer's disease. *The American Journal of Medicine*, *103*(3), 3S–10S. https://doi.org/10.1016/S0002-9343(97)00262-3

Celardo, I., Pedersen, J. Z., Traversa, E., & Ghibelli, L. (2011). Pharmacological potential of cerium oxide nanoparticles. *Nanoscale*, *3*(4), 1411. https://doi.org/10.1039/c0nr00875c

Chu, J., Qin, R., Wang, S.-J., Wang, Q., & Wu, Q. (2025). Integrated single-cell and transcriptomic analysis of bone marrow-derived metastatic neuroblastoma reveals molecular mechanisms of metabolic reprogramming. *Scientific Reports, 15*(1), 28519. https://doi.org/10.1038/s41598-025-13626-8

Costa, C. P., Moreira, J. N., Sousa Lobo, J. M., & Silva, A. C. (2021). Intranasal delivery of nanostructured lipid carriers, solid lipid nanoparticles and nanoemulsions: A current overview of in vivo studies. *Acta Pharmaceutica Sinica B, 11*(4), 925–940. https://doi.org/10.1016/j.apsb.2021.02.012

Cui, M. Y., Lin, Y., Sheng, J. Y., Zhang, X., & Cui, R. J. (2018). Exercise intervention associated with cognitive improvement in Alzheimer's disease. *Neural Plasticity, 2018*, 1–10. https://doi.org/10.1155/2018/9234105

Cummings, J. L., Morstorf, T., & Zhong, K. (2014). Alzheimer's disease drug-development pipeline: Few candidates, frequent failures. *Alzheimer's Research & Therapy, 6*(4), 37. https://doi.org/10.1186/alzrt269

Cummings, J., Ritter, A., & Zhong, K. (2018). Clinical trials for disease-modifying therapies in Alzheimer's disease: A primer, lessons learned, and a blueprint for the future. *Journal of Alzheimer's Disease, 64*(s1), S3–S22. https://doi.org/10.3233/JAD-179901

Dara, T., Vatanara, A., Sharifzadeh, M., Khani, S., Vakilinezhad, M. A., Vakhshiteh, F., Nabi Meybodi, M., Sadegh Malvajerd, S., Hassani, S., & Mosaddegh, M. H. (2019). Improvement of memory deficits in the rat model of Alzheimer's disease by erythropoietin-loaded solid lipid nanoparticles. *Neurobiology of Learning and Memory, 166*, 107082. https://doi.org/10.1016/j.nlm.2019.107082

Davoody, S., Asgari Taei, A., Khodabakhsh, P., & Dargahi, L. (2024). mTOR signaling and Alzheimer's disease: What we know and where we are? *CNS Neuroscience & Therapeutics, 30*(4). https://doi.org/10.1111/cns.14463

Deardorff, W. J., & Grossberg, G. (2016). A fixed-dose combination of memantine extended-release and donepezil in the treatment of moderate-to-severe Alzheimer's disease. *Drug Design, Development and Therapy, 10*, 3267–3279. https://doi.org/10.2147/DDDT.S86463

El Kadmiri, N., Said, N., Slassi, I., El Moutawakil, B., & Nadifi, S. (2018). Biomarkers for Alzheimer disease: Classical and novel candidates' review. *Neuroscience, 370*, 181–190. https://doi.org/10.1016/j.neuroscience.2017.07.017

Evans, R. C., Dar, N. J., Chen, L., Na, R., O'Connor, J. C., Jiang, J., Zheng, S., & Ran, Q. (2025). Inhibition of Hippo signaling through ablation of Lats1 and Lats2 protects against cognitive decline in 5xFAD mice via increasing neuronal resilience against ferroptosis. *Aging Cell, 24*(11). https://doi.org/10.1111/acel.70218

Fonte, J., Miklossy, J., Atwood, C., & Martins, R. (2001). The severity of cortical Alzheimer's type changes is positively correlated with increased amyloid-β levels: Resolubilization of amyloid-β with transition metal ion chelators. *Journal of Alzheimer's Disease, 3*(2), 209–219. https://doi.org/10.3233/JAD-2001-3206

Gajendra, K., Pratap, G. K., Poornima, D. V., Shantaram, M., & Ranjita, G. (2024). Natural acetylcholinesterase inhibitors: A multi-targeted therapeutic potential in Alzheimer's disease. *European Journal of Medicinal Chemistry Reports, 11*, 100154. https://doi.org/10.1016/j.ejmcr.2024.100154

Galle, S. A., Geraedts, I. K., Deijen, J. B., Milders, M. V., & Drent, M. L. (2020). The interrelationship between insulin-like growth factor 1, apolipoprotein E ε4, lifestyle factors, and the aging body and brain. *The Journal of Prevention of Alzheimer's Disease, 7*(4), 265–273. https://doi.org/10.14283/jpad.2020.11

Galvan, V., Brandimarti, R., Munger, J., & Roizman, B. (2000). Bcl-2 blocks a caspase-dependent pathway of apoptosis activated by herpes simplex virus 1 infection in HEp-2 cells. *Journal of Virology, 74*(4), 1931–1938. https://doi.org/10.1128/JVI.74.4.1931-1938.2000

Ganguly, S. C., Sangram, S., Paul, S., & Kundu, M. (2024). Phyto-nanotechnology: A novel beneficial strategy for Alzheimer's disease therapy. *Neurochemistry International, 180*, 105868. https://doi.org/10.1016/j.neuint.2024.105868

Gao, C., Jiang, J., Tan, Y., & Chen, S. (2023). Microglia in neurodegenerative diseases: Mechanism and potential therapeutic targets. *Signal Transduction and Targeted Therapy, 8*(1), 359. https://doi.org/10.1038/s41392-023-01588-0

Gobbi, M., Re, F., Canovi, M., Beeg, M., Gregori, M., Sesana, S., Sonnino, S., Brogioli, D., Musicanti, C., Gasco, P., Salmona, M., & Masserini, M. E. (2010). Lipid-based nanoparticles with high binding affinity for amyloid-β1–42 peptide. *Biomaterials, 31*(25), 6519–6529. https://doi.org/10.1016/j.biomaterials.2010.04.044

Goudarzi, N., Taheri, Z., Nezhad Salari, A. M., Kazemzadeh, K., & Tafakhori, A. (2025). Recognition and classification of facial expression using artificial intelligence as a key of early detection in neurological disorders. *Reviews in the Neurosciences, 36*(5), 479–495. https://doi.org/10.1515/revneuro-2024-0125

Gu, Z., Ge, B., Wang, Y., Gong, Y., & Qi, M. (2025). Artificial intelligence technologies for enhancing neurofunctionalities: A comprehensive review with applications in Alzheimer's disease research. *Frontiers in Aging Neuroscience, 17.* https://doi.org/10.3389/fnagi.2025.1609063

Guo, S., Wang, H., & Yin, Y. (2022). Microglia polarization from M1 to M2 in neurodegenerative diseases. *Frontiers in Aging Neuroscience, 14*, https://doi.org/10.3389/fnagi.2022.815347

Gupta, J., Fatima, M. T., Islam, Z., Khan, R. H., Uversky, V. N., & Salahuddin, P. (2019). Nanoparticle formulations in the diagnosis and therapy of Alzheimer's disease. *International Journal of Biological Macromolecules, 130*, 515–526. https://doi.org/10.1016/j.ijbiomac.2019.02.156

Han, X., Meng, X., Wu, Y., Xia, W., Xue, S., Liu, X., Lyu, C., Li, Z., Yan, X., Won Jung, H., & Zhang, S. (2025). Systems pharmacology identifies ajugol-mediated NF-κB/caspase-3 inhibition and isoacteoside-driven p62/mTOR-mediated autophagy as key mechanisms of Rehmanniae Radix and its processed form in Alzheimer's treatment. *Frontiers in Pharmacology, 16*, https://doi.org/10.3389/fphar.2025.1644847

Hansson, O. (2021). Biomarkers for neurodegenerative diseases. *Nature Medicine, 27*(6), 954–963. https://doi.org/10.1038/s41591-021-01382-x

Hein, Z. M., Vishnumukkala, T., Karikalan, B., Alkatiri, A., Hussan, F., Jagadeesan, S., Kamaruzzaman, M. A., Che Ramli, M. D., Che Mohd Nassir, C. M. N., & Gopalakrishna, P. K. (2025). Autophagy and Alzheimer's disease: Mechanisms and impact beyond the brain. *Cells, 14*(12), 911. https://doi.org/10.3390/cells14120911

Ibrahim, M. M., & Gabr, M. T. (2019). Multitarget therapeutic strategies for Alzheimer's disease. *Neural Regeneration Research, 14*(3), 437. https://doi.org/10.4103/1673-5374.245463

Ismaili, L., Refouvelet, B., Benchekroun, M., Brogi, S., Brindisi, M., Gemma, S., Campiani, G., Filipic, S., Agbaba, D., Esteban, G., Unzeta, M., Nikolic, K., Butini, S., & Marco-Contelles, J. (2017). Multitarget compounds bearing tacrine- and donepezil-like structural and functional motifs for the potential treatment of Alzheimer's disease. *Progress in Neurobiology, 151*, 4–34. https://doi.org/10.1016/j.pneurobio.2015.12.003

Kale, M., Wankhede, N., Pawar, R., Ballal, S., Kumawat, R., Goswami, M., Khalid, M., Taksande, B., Upaganlawar, A., Umekar, M., Kopalli, S. R., & Koppula, S. (2024). AI-driven innovations in Alzheimer's disease: Integrating early diagnosis, personalized treatment, and prognostic modelling. *Ageing Research Reviews, 101*, 102497. https://doi.org/10.1016/j.arr.2024.102497

Kwon, H. J., Cha, M.-Y., Kim, D., Kim, D. K., Soh, M., Shin, K., Hyeon, T., & Mook-Jung, I. (2016). Mitochondria-targeting ceria nanoparticles as antioxidants for Alzheimer's disease. *ACS Nano, 10*(2), 2860–2870. https://doi.org/10.1021/acsnano.5b08045

Li, Q., Yang, H., Chen, Y., & Sun, H. (2017). Recent progress in the identification of selective butyrylcholinesterase inhibitors for Alzheimer's disease. *European Journal of Medicinal Chemistry, 132*, 294–309. https://doi.org/10.1016/j.ejmech.2017.03.062

Li, X., Wu, Z., Si, X., Li, J., Wu, G., & Wang, M. (2025). The role of mitochondrial dysfunction in the pathogenesis of Alzheimer's disease and future strategies for targeted therapy. *European Journal of Medical Research, 30*(1), 434. https://doi.org/10.1186/s40001-025-02699-w

Lin, J., Yuan, Y., Huang, C., Zi, J., Li, L., Liu, J., Wu, X., Li, W., Zhao, Q., Li, Y., Liu, Z., & Diao, A. (2024). TFEB agonist clomiphene citrate activates the autophagy-lysosomal pathway and ameliorates Alzheimer's disease symptoms in mice. *Journal of Biological Chemistry*, *300*(12), 107929. https://doi.org/10.1016/j.jbc.2024.107929

Liu, G., Men, P., Harris, P. L. R., Rolston, R. K., Perry, G., & Smith, M. A. (2006). Nanoparticle iron chelators: A new therapeutic approach in Alzheimer disease and other neurologic disorders associated with trace metal imbalance. *Neuroscience Letters*, *406*(3), 189–193. https://doi.org/10.1016/j.neulet.2006.07.020

Liu, S., Jin, X., Ge, Y., Dong, J., Liu, X., Pei, X., Wang, P., Wang, B., Chang, Y., & Yu, X. (2025). Advances in brain-targeted delivery strategies and natural product-mediated enhancement of blood–brain barrier permeability. *Journal of Nanobiotechnology*, *23*(1), 382. https://doi.org/10.1186/s12951-025-03415-w

Luo, W., Bu, W., Zhang, G., Dong, Y., Wang, Y., Wang, J., Liu, C., Hu, X., Jia, Y., & Ren, H. (2025). Downregulation of Nrf2 deteriorates cognitive impairment in APP/PS1 mice by inhibiting mitochondrial biogenesis through the PPARγ/PGC1α signaling pathway. *Behavioural Brain Research*, *495*, 115805. https://doi.org/10.1016/j.bbr.2025.115805

Manubolu, K., & Peeriga, R. (2025). Revolutionizing neurodegenerative disease management: The synergy of AI and pharmacy. *Current Aging Science*, *18*, https://doi.org/10.2174/0118746098375978250820220024

Mathew, A., Fukuda, T., Nagaoka, Y., Hasumura, T., Morimoto, H., Yoshida, Y., Maekawa, T., Venugopal, K., & Kumar, D. S. (2012). Curcumin loaded-PLGA nanoparticles conjugated with Tet-1 peptide for potential use in Alzheimer's disease. *PLoS ONE*, *7*(3), e32616. https://doi.org/10.1371/journal.pone.0032616

Meng, Q., Wang, A., Hua, H., Jiang, Y., Wang, Y., Mu, H., Wu, Z., & Sun, K. (2018). Intranasal delivery of Huperzine A to the brain using lactoferrin-conjugated N-trimethylated chitosan surface-modified PLGA nanoparticles for treatment of Alzheimer's disease. *International Journal of Nanomedicine*, *13*, 705–718. https://doi.org/10.2147/IJN.S151474

Miao, J., Ma, H., Yang, Y., Liao, Y., Lin, C., Zheng, J., Yu, M., & Lan, J. (2023). Microglia in Alzheimer's disease: Pathogenesis, mechanisms, and therapeutic potentials. *Frontiers in Aging Neuroscience*, *15*, https://doi.org/10.3389/fnagi.2023.1201982

Mir Najib Ullah, S. N., Afzal, O., Altamimi, A. S. A., Ather, H., Sultana, S., Almalki, W. H., Bharti, P., Sahoo, A., Dwivedi, K., Khan, G., Sultana, S., Alzahrani, A., & Rahman, M. (2023). Nanomedicine in the management of Alzheimer's disease: State-of-the-art. *Biomedicines*, *11*(6), 1752. https://doi.org/10.3390/biomedicines11061752

Monteiro, A. R., Barbosa, D. J., Remião, F., & Silva, R. (2023). Alzheimer's disease: Insights and new prospects in disease pathophysiology, biomarkers and disease-modifying drugs. *Biochemical Pharmacology*, *211*, 115522. https://doi.org/10.1016/j.bcp.2023.115522

Nani, A., Murtaza, B., Sayed Khan, A., Khan, N. A., & Hichami, A. (2021). Antioxidant and anti-inflammatory potential of polyphenols contained in Mediterranean diet in obesity: Molecular mechanisms. *Molecules*, *26*(4), 985. https://doi.org/10.3390/molecules26040985

Olczak, K. J., Taylor-Bateman, V., Nicholls, H. L., Traylor, M., Cabrera, C. P., & Munroe, P. B. (2021). Hypertension genetics past, present and future applications. *Journal of Internal Medicine*, *290*(6), 1130–1152. https://doi.org/10.1111/joim.13352

Panwar, D., Sharma, P., Sharma, S., Goyal, M., & Goyal, K. (2024). *Role of artificial intelligence in cognitive assessment and early detection of Alzheimer's disease*, 190–210. https://doi.org/10.4018/979-8-3693-3605-2.ch013

Plaza-Oliver, M., Santander-Ortega, M. J., & Lozano, M. V. (2021). Current approaches in lipid-based nanocarriers for oral drug delivery. *Drug Delivery and Translational Research*, *11*(2), 471–497. https://doi.org/10.1007/s13346-021-00908-7

Rahman, Md. A., Rahman, M. D. H., Biswas, P., Hossain, Md. S., Islam, R., Hannan, Md. A., Uddin, M. J., & Rhim, H. (2020). Potential therapeutic role of phytochemicals to mitigate mitochondrial dysfunctions in Alzheimer's disease. *Antioxidants*, *10*(1), 23. https://doi.org/10.3390/antiox10010023

Ramonet, D., Daerr, A., & Hallbeck, M. (2025). Stabilizing the retromer complex rescues synaptic dysfunction and endosomal trafficking deficits in an Alzheimer's disease mouse model. *Acta Neuropathologica Communications*, *13*(1), 190. https://doi.org/10.1186/s40478-025-02096-8

Reardon, S. (2023). FDA approves Alzheimer's drug lecanemab amid safety concerns. *Nature*, *613*(7943), 227–228. https://doi.org/10.1038/d41586-023-00030-3

Sadiq, I. Z. (2023). Free radicals and oxidative stress: signaling mechanisms, redox basis for human diseases, and cell cycle regulation. *Current Molecular Medicine*, *23*(1), 13–35. https://doi.org/10.2174/1566524022666211222161637

Saptarshi, N., Porter, L. F., & Paraoan, L. (2022). PERK/EIF2AK3 integrates endoplasmic reticulum stress-induced apoptosis, oxidative stress and autophagy responses in immortalised retinal pigment epithelial cells. *Scientific Reports*, *12*(1), 13324. https://doi.org/10.1038/s41598-022-16909-6

Savelieff, M. G., Nam, G., Kang, J., Lee, H. J., Lee, M., & Lim, M. H. (2019). Development of multifunctional molecules as potential therapeutic candidates for Alzheimer's disease, Parkinson's disease, and amyotrophic lateral sclerosis in the last decade. *Chemical Reviews*, *119*(2), 1221–1322. https://doi.org/10.1021/acs.chemrev.8b00138

Schneider, L. S., Mangialasche, F., Andreasen, N., Feldman, H., Giacobini, E., Jones, R., Mantua, V., Mecocci, P., Pani, L., Winblad, B., & Kivipelto, M. (2014). Clinical trials and late-stage drug development for Alzheimer's disease: An appraisal from 1984 to 2014. *Journal of Internal Medicine*, *275*(3), 251–283. https://doi.org/10.1111/joim.12191

Se Thoe, E., Fauzi, A., Tang, Y. Q., Chamyuang, S., & Chia, A. Y. Y. (2021). A review on advances of treatment modalities for Alzheimer's disease. *Life Sciences*, *276*, 119129. https://doi.org/10.1016/j.lfs.2021.119129

Shanmugam, J. V., Duraisamy, B., Simon, B. C., & Bhaskaran, P. (2022). Alzheimer's disease classification using pre-trained deep networks. *Biomedical Signal Processing and Control*, *71*, 103217. https://doi.org/10.1016/j.bspc.2021.103217

Sharma, A., Rudrawar, S., Bharate, S. B., & Jadhav, H. R. (2025). Recent advancements in the therapeutic approaches for Alzheimer's disease treatment: Current and future perspective. *RSC Medicinal Chemistry*, *16*(2), 652–693. https://doi.org/10.1039/D4MD00630E

Sheikh, K. A., Muzammil, S., Raneem, E., Akhter, M., Anwer, T., Meenu, C. S., Kaleem, M., Saifullah, Md. K., Alam, M. M., & Shaquiquzzaman, M. (2025). Tranylcypromine-derived LSD1 inhibitors: Synthetic strategies, structure-activity relationships, and anticancer potential. *European Journal of Medicinal Chemistry*, *299*, 118054. https://doi.org/10.1016/j.ejmech.2025.118054

Shi, D., Guo, X., Ning, Z., Zhang, Y., Liu, F., Liu, M., & Wei, Y. (2025). Review of FUNDC1-mediated mitochondrial autophagy in Alzheimer's disease. *Frontiers in Aging Neuroscience*, *17*, https://doi.org/10.3389/fnagi.2025.1544241

Soldevila-Matías, P., Sánchez-Ortí, J. V., Correa-Ghisays, P., Balanzá-Martínez, V., Selva-Vera, G., Sanchis-Sanchis, R., Iglesias-García, N., Monfort-Pañego, M., Tomás-Martínez, P., Victor, V. M., Crespo-Facorro, B., San-Martín Valenzuela, C., Climent Sánchez, J. A., Corral-Márquez, R., & Tabarés-Seisdedos, R. (2024). Exercise as a promoter of neurocognitive improvement in people with psychiatric disorders and comorbid obesity: A randomized controlled trial. *Psychiatry Research*, *342*, 116226. https://doi.org/10.1016/j.psychres.2024.116226

Tabarestani, S., Aghili, M., Eslami, M., Cabrerizo, M., Barreto, A., Rishe, N., Curiel, R. E., Loewenstein, D., Duara, R., & Adjouadi, M. (2020). A distributed multitask multimodal approach for the prediction of Alzheimer's disease in a longitudinal study. *NeuroImage*, *206*, 116317. https://doi.org/10.1016/j.neuroimage.2019.116317

Torrecillas-Lopez, M., Claro-Cala, C. M., Gonzalez-de la Rosa, T., Barrera-Chamorro, L., Millan-Linares, M. C., Marquez-Paradas, E., Villanueva, A., del Rio-Vazquez, J. L., & Montserrat-de la Paz, S. (2025). Neuroavailable peptides from hempseed protein hydrolysates reduce hippocampal inflammation and glial activation in a scopolamine-induced Alzheimer's disease. *Biomedicine & Pharmacotherapy*, *191*, 118438. https://doi.org/10.1016/j.biopha.2025.118438

Vilella, A., Belletti, D., Sauer, A. K., Hagmeyer, S., Sarowar, T., Masoni, M., Stasiak, N., Mulvihill, J. J. E., Ruozi, B., Forni, F., Vandelli, M. A., Tosi, G., Zoli, M., & Grabrucker, A. M. (2018). Reduced plaque size and inflammation in the APP23 mouse model for Alzheimer's disease after chronic application of polymeric nanoparticles for CNS targeted zinc delivery. *Journal of Trace Elements in Medicine and Biology*, *49*, 210–221. https://doi.org/10.1016/j.jtemb.2017.12.006

Wang, H., Pan, J., Zhang, M., & Tan, Z. (2025). Re-evaluation of the efficacy and safety of anti-Aβ monoclonal antibodies (lecanemab/donanemab) in the treatment of early Alzheimer's disease. *Frontiers in Pharmacology*, *16*, https://doi.org/10.3389/fphar.2025.1599048

Weller, J., & Budson, A. (2018). Current understanding of Alzheimer's disease diagnosis and treatment. *F1000Research*, *7*, 1161. https://doi.org/10.12688/f1000research.14506.1

Wu, A., Lee, D., & Xiong, W.-C. (2024). VPS35 or retromer as a potential target for neurodegenerative disorders: Barriers to progress. *Expert Opinion on Therapeutic Targets*, *28*(8), 701–712. https://doi.org/10.1080/14728222.2024.2392700

Wu, W., Ji, Y., Wang, Z., Wu, X., Li, J., Gu, F., Chen, Z., & Wang, Z. (2023). The FDA-approved anti-amyloid-β monoclonal antibodies for the treatment of Alzheimer's disease: A systematic review and meta-analysis of randomized controlled trials. *European Journal of Medical Research*, *28*(1), 544. https://doi.org/10.1186/s40001-023-01512-w

Xu, J., Shen, R., Qian, M., Zhou, Z., Xie, B., Jiang, Y., Yu, Y., & Dong, W. (2025). Ferroptosis in Alzheimer's disease: The regulatory role of glial cells. *Journal of Integrative Neuroscience*, *24*(4). https://doi.org/10.31083/JIN25845

Xu, W., Geng, Y., Liu, W., Fan, X., & Zhang, N. (2025). *ADInformer for the personalized prediction of Alzheimer's disease progression trajectory* (pp. 309–320). https://doi.org/10.1007/978-981-96-9986-5_26

Zeng, H., & Jin, Z. (2025). The role of ferroptosis in Alzheimer's disease: Mechanisms and therapeutic potential (review). *Molecular Medicine Reports*, *32*(1), 1–10. https://doi.org/10.3892/mmr.2025.13557

Zetterberg, H., & Bendlin, B. B. (2021). Biomarkers for Alzheimer's disease—preparing for a new era of disease-modifying therapies. *Molecular Psychiatry*, *26*(1), 296–308. https://doi.org/10.1038/s41380-020-0721-9

Zhang, Y., Liu, T., Lanfranchi, V., & Yang, P. (2023). Explainable tensor multi-task ensemble learning based on brain structure variation for Alzheimer's disease dynamic prediction. *IEEE Journal of Translational Engineering in Health and Medicine*, *11*, 1–12. https://doi.org/10.1109/JTEHM.2022.3219775

Zhou, Z., An, Q., Zhang, W., Li, Y., Zhang, Q., & Yan, H. (2024). Histamine and receptors in neuroinflammation: Their roles on neurodegenerative diseases. *Behavioural Brain Research*, *465*, 114964. https://doi.org/10.1016/j.bbr.2024.114964

MIX
Papier aus verantwortungsvollen Quellen
Paper from responsible sources
FSC® C105338

If you have any concerns about our products,
you can contact us on
ProductSafety@springernature.com

In case Publisher is established outside the EU,
the EU authorized representative is:
Springer Nature Customer Service Center GmbH
Europaplatz 3, 69115 Heidelberg, Germany

Printed by Libri Plureos GmbH
in Hamburg, Germany